T0094051

THE LITTLE BOOK

of *bees*

THE LITTLE BOOK

of ***bees***

KARL WEISS

with Carlos H. Vergara

COPERNICUS BOOKS

An Imprint of Springer-Verlag

Originally published as *Bienen und Bienenvölker,*
© 1997 Verlag C. H. Beck oHG, München, Germany.

© 2003 Springer-Verlag New York, Inc.
Softcover reprint of the hardcover 1st edition 2003

Published in the United States by Copernicus Books,
an imprint of Springer-Verlag New York, Inc.
A member of BertelsmannSpringer Science+Business Media GmbH

Copernicus Books
37 East 7th Street
New York, NY 10003
www.copernicusbooks.com

Library of Congress Cataloging-in-Publication Data
Weiss, Karl.
    [Bienen und Bienenvölker. English]
    The little book of bees / Karl Weiss, Carlos H. Vergara.
        p. cm. — (Little book series)
    Includes bibliographical references and index.

    1. Bees—Behavior. I. Vergara, Carlos H. II. Title. III. Little book
    series (New York, N.Y.)
    QL563.W4513 2002
    595.79'9—dc21   2002067537

Manufactured in the United States of America.
Printed on acid-free paper.
Translated by Douglas Haynes.

9 8 7 6 5 4 3 2 1

ISBN 978-1-4419-2922-8 SPIN 10795940

Preface   vii

chapter 1 Bees in the Animal Kingdom
             and in Earth's History   1

chapter 2 The Bee—An Insect   15

chapter 3 What Does "Social"
             Mean in the Animal Kingdom?   25
             Anonymous Societies   26
             Family Associations   28
             The Insect Colony   30
             Social Classifications   34

chapter 4 The Traits of Bees and Their System   35

chapter 5 Solitary Bees and Social Development   45
             Solitary Bees   45
             On the Way to a Colony   63
             Unique Mating Behavior   75
             Threats from Outside, and Enemies Within the Ranks   77

chapter 6 Bumblebees and Stingless Bees   83
             Bumblebees   84
             Stingless Bees   97

chapter 7 On the Summit of Social Insect Life   107

The Genus *Apis*: Species and Races   108

The Domestic Honeybee   118

The Comb   118

The Colony and Its Individuals   122

Brood Rearing and Division of Labor   126

What Holds a Colony Together   128

How New Colonies Originate   130

Sensory Capabilities   132

Artists of Orientation   136

The Language of Honeybees   138

chapter 8 Nest Aids for Wild Bees   145

Abridged Bibliography
and Further Reading   151

Index   155

Preface

People generally think of bees as insects that fly out of hives and make us honey. Actually, there are many different kinds of bees, most of which live relatively hidden from our view. Maybe some of us are familiar with bumblebees, but who would recognize the solitary bees, which considerably outnumber the social bees?

These little-observed bees are often dubbed "wild bees" to distinguish them from our honeymakers. But fundamentally, our honeybees are not domesticated either, no matter how much we've tried over centuries to make them so. They have persistently retained the life skills nature gave them. Left to their own devices in the wild, they will easily survive and reproduce if they find a suitable nesting place, such as a hollow tree or a protected hole in the ground. Such bees, which often escape from apiaries as swarms, should, strictly speaking, be called wild. But this is not customary, so we won't refer to them that way, either. The best way for us to define wild bees is to simply consider all bees that don't produce honey for people as wild.

Our honeybee colonies are highly developed social entities with community characteristics we can hardly imagine. These complex social structures emerged out of simpler preliminary arrangements over long periods of Earth's history, as did all forms of the plant and animal kingdoms. Without a doubt, solitary bees preceded them. By investigating the evolutionary history of bees, we can get a sense of the wondrous variety of bees and bees' social structures.

This *Little Book* is not directed at specialists. Instead, it's intended for everyone who is interested in the solitary and social lives of bees. Although the text does not go beyond its prescribed scope as a *Little Book*, I've made every effort to include all of the details necessary to understand bees in context. To improve readability, the text has not been burdened with footnotes.

If, at the end of the book, you feel compelled to get acquainted with the world of bees through your own eyes by creating nests for solitary bees or bumblebees, or even a honeybee hive, you will find help and encouragement in the final chapter and in designated titles in the bibliography. This book aims not just to teach readers, but encourage them to participate—and that would be the author's greatest reward.

Karl Weiss

chapter 1

Bees in the Animal Kingdom and in Earth's History

In order to gain an overview of our planet's animal diversity, we humans have developed a scientific classification system, in which every life form holds an allotted place. Classification occurs according to universally applicable, internationally recognized rules established by the Swedish naturalist Carolus Linnaeus in his famous work *Systema Naturae*, published in 1735. Since then, all known living creatures have been divided into categories according to their different physiological and behavioral characteristics. The basis of this system is the concept of *species*. Animals (or plants) that share or at least closely resemble each other in appearance and traits and can produce offspring together belong to a single species. If two organisms can't produce offspring together, then they are members of two different species, even if they appear to be very similar. Differing characteristics within a species (frequently due to geography) lead to so-called subspecies, or races. Similar species are grouped in a genus. Linnaeus gave every organism two names: a genus name and a species name. This is still done today. The names are taken from Latin or are Latin imitations, which allows scientists all over the world to understand them. The genus name begins with a capital letter, and the species name begins with a small letter. The subspecies name follows thereafter (typically called a race in honeybees), and the name of the discoverer or first person to describe the species, with the year of discovery, comes last (in especially precise sources). For example, one of the more well-known honeybees among beekeepers in central Europe is called the Italian bee. Its scientific name is *Apis mellifera ligustica* SPINOLA (1908). *Apis* is the genus, *mellifera* is the species, *ligustica* is the race, and SPINOLA was the first person to describe this race.

Linnaeus's system puts groups of genera (more than one genus) together to create further classifications (in order of smallest to largest): family, order, class, and phylum. Since Linnaeus's time this helpful system has been gradually expanded. Taxonomists are continually occupied with improving and refining this huge system for organizing the living world.

Among the numerous phyla classified in modern times are some of the animal groups most familiar to lay people: sponges, coelenterates (jellyfish, corals, sea anemones), jointed worms, echinoderms (starfish, sea urchins), mollusks, arthropods (insects, spiders, crabs, mites), and chordates, which include the well-known subphylum

Vertebrata and its classes fish, amphibians, reptiles, birds, and mammals. Bees belong to the phylum *Arthropoda*, and within that phylum to the class of insects. This class contains by far the most species in the entire animal kingdom. While the vertebrate classes together include only about 50,000 species, well over 1 million insect species have been described up to the present. Within the class of insects bees belong to the order *Hymenoptera*. Among insects, only Hymenoptera and the order of termites (*Isoptera*) have developed true animal societies.

Although termites are not the focus of this book, you should know that the 2000 species of termites, which have much less complex body structures and developmental stages than Hymenoptera, can build nests containing multiple millions of individuals. The size of termites' nests is perhaps only surpassed by the nests of tropical army ants (*Dorylina*), which include up to 20 million "marchers." Among ants and termites, there are no longer any solitary species.

The more than 100,000 species of the order Hymenoptera, the largest insect order, are subdivided into three groups (superfamilies): *Vespoidea* (yellow jackets, wasps), *Apoidea* (bees), and *Formicoidea* (ants). In contrast to ants and termites, most wasp and bee species are solitary, with the remainder ranging from slightly social to definitively social.

Figure 1.1 shows how the order Hymenoptera is subdivided, and where bees are located within the order. It provides a general impression of the diversity among

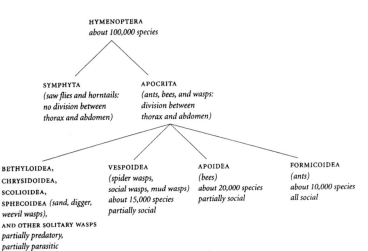

FIGURE 1.1

Overview of the order Hymenoptera with information on a few prominent families and family groups, respectively.

Hymenoptera and positions the bees as one of many selected superfamilies.

Paleontologists seek information that tells us how life forms originated over the course of Earth's history. With this information, they can draw conclusions about the family relationships of living organisms. Fossils are especially helpful for this task because the geological deposits they are found in reveal roughly how old they are.

Even without fossils, we believe we can make statements about the bees' first appearance on Earth. One of the bees' peculiarities, and one of the definite differences between them and most of their Hymenoptera relatives, is their diet. While ants and wasps require supplemental animal food to nourish themselves, bees are purely herbivorous. They prefer the sweet saps offered primarily by flowering plants in the form of nectar and especially need pollen as an indispensable source of protein. Therefore bees depend on the flowering plants (angiosperms) for their food supply, which leads us to conclude that the first bees couldn't have existed before the appearance of this plant type. Flowering plants first appeared in the mid-Cretaceous period, about 100 million years ago. Although there have been no discoveries of fossilized bees earlier than this, it's nevertheless not obvious that flowering plants preceded bees. Forerunners of bees could have nourished themselves with the produce of gymnosperms (plants, such as ginkgos, that have seeds unprotected by ovaries), the spores of ferns, and pollen from wind-pollinated conifers, all of which existed long before flowering plants. The impetus for the development of flowers in angiosperms could just as well have been initiated by bees. Once bees appeared with their covering of hairs—a transmitter of pollen—the plants could have adjusted to them and begun to attract them with sweet secretions and all kinds of subtle odors and colors. Led by their desire for sweetness, the bees would have responded with growing resourcefulness and readiness to learn. Consequently, some flowering plants might have stored their nectar deeper and developed calyx tubes and nectar spurs. In this way, they might have induced certain bees to develop longer, more functional proboscises that allowed them to attach to these flowers, in particular. At any rate, what's certain is that bees and flowering plants reciprocally influenced each other's development and are dependent on each other today. For them, we have the sensible term coevolution (from the Latin *co*, "together," and *evolutio*, "development").

Plant resin was an ideal medium for preserving small primeval life forms, particularly insects. Over millions of years it hardened into amber. We find amber in geological deposits from the Cretaceous and Tertiary periods of Earth's history—everywhere coniferous forests produced above-average flows of resin. In an upper-Cretaceous amber deposit in New Jersey, American entomologists have found the oldest fossilized bee yet discovered, approximately 90 million years old. It may have already been social then, and it resembles stingless bees now living in tropical and subtropical regions. The Baltic Sea coast, with its Baltic amber deposits from the early Tertiary (the Eocene, about 56 million years ago), is a rich storehouse of primeval bees.

These specimens, later described under the name *Electrapis*, also show the characteristics of stingless bees, and they resemble bumblebees and honeybees, as well.

A similar find made recently at a fossil site in the once-volcanic Eifel Mountains of Germany was dated as mid-Eocene, about 45 million years ago. A crater lake formed there in a funnel created by a volcanic eruption at the beginning of the Tertiary. The rain in the then subtropical climate eroded fine matter from the rim of the crater into the lake. Here, various fossils of plant and animal origin are deposited in the finely layered clays. Among them are uncounted numbers of insects, including the bee we have enthusiastically (and certainly somewhat prematurely) dubbed "the oldest honeybee in the world." Up until now, the oldest fossils of true honeybees with undisputed colony-building characteristics were found in the Seven Mountains near Bonn, Germany. These fossils appeared in leaves of coal from the lower Miocene period, approximately 23 million years ago. The coal comes from plant and animal debris that thickened in the oxygen-poor bottom of a gradually silting-in freshwater lake. As the coal, streaked with bright sand and gravel layers, was mined, a wealth of fossilized life was discovered, including many different kinds of insects and the well-preserved honeybees. Other finds of fossil bees have come from coal deposits in the Randecker Crater in Swabia, a region in southwestern Germany. These somewhat younger fossils, from the upper Miocene about 12 million years ago, almost exactly match the size and appearance of today's living bee species *Apis mellifera*. The same can be said of other bee fossils of about the same age found in the iron-drossed limestone layers of a spring in Böttingen (also in Swabia).

Nevertheless, the honeybees living in Europe today are probably not direct descendants of the bees that lived there under tropical and subtropical climatic conditions in the Tertiary. Along with many other heat-loving plants and animals, the Tertiary bees likely abandoned Europe well before the beginning of the Quaternary Ice Age. Our contemporary "western honeybee" probably has its roots in South and Southwest Asia, where it departed for Europe and Africa beginning 1 to 2 million years ago. Asia is still the home of all honeybee species today.

The phylogenetic predecessors of bees are wasp-like organisms, probably solitary digger wasps of the family group *Sphecidae*, which feed their broods captured arthropods such as spiders, beetles, flies, bees, true bugs, butterflies, or caterpillars. As a rule, each species of these wasps is committed to certain prey. They paralyze their victims using a sting, carry them to a hiding-place they've excavated in the ground, and lay their egg in the supply of flesh (the prey). Often, they close off these nests right away. In some species, the female remains after laying her egg to take care of the growing larvae, supplying them with food. Such behavior is displayed by the many representatives of the digger wasps alive today: sand wasps, bembicinine sand wasps, mellininine wasps, weevil wasps, spiny digger wasps, potter wasps, and bee wolves.

We estimate that the first appearance of the first digger wasps was between the Jurassic and Cretaceous, about 145 million years ago—not long after the first true birds raised themselves into the air. A hundred million years before that (in the Triassic), we can assume the Hymenoptera appeared. Their first representatives were the *Symphyta*, who had no division between thorax and abdomen, like today's saw flies and wood wasps.

The first ants, initially probably strictly pedestrian (without wings), could have emerged on Earth somewhat earlier than the bees. We presume, however, that they only appeared after the first wasp-like organisms, from which we think they came. The oldest ant fossils were found in resin from the upper Cretaceous in New Jersey. They are about 90 million years old and show the stipellus between thorax and abdomen typical of ants, though it is short like in wasps. They also have two teeth against the mandible. Younger ant-like forms from the lower Cretaceous (about 35 million years ago) have been found in Lebanon, Australia, and Canada.

The oldest colony-building insects are the termites (Isoptera). Their many representatives are still somewhat primeval-looking today. With their waistless bodies, identical pairs of legs, and identical simple-veined front and hind wings, they resemble their forerunners, cockroaches. The oldest uncontested termite find comes from Labrador from the upper Cretaceous, not any older than finds of wasp-like Hymenoptera. These *Cretatermes*, of the family *Hodotermi-*

tidae, are believed to be 100 million years old. Since these termites had already developed colony-building characteristics, their discoverer Emerson believes they diverged from their cockroach ancestors much earlier, as far back as the late Paleozoic. Fossil finds of cockroaches, as well as beetles and spiders, show they lived as far back as the Permian, approximately 250 million years ago.

As a class, insects are quite a bit older yet. We date their origin between the Devonian and Carboniferous periods, before the appearance of reptiles. Insects abandoned the water at about the same time amphibians first ventured onto land. The first insects to do so were certainly the wingless *Apterygota* and *Collembola* (springtails), the latter of which have been found in strata deposited in the middle Devonian. The wings they developed (*Pterygota*) were surely rigid at first and possibly only useful for gliding. This stage was surpassed in the Carboniferous, at the latest, when there were dragonfly-type insects 25 centimeters long with a wingspan of half a meter. Whether insects evolved from jointed worms or crab-like creatures continues to be contested.

When we compare the origin of bees, probably the youngest colony-building insect, with the first appearance of humans, it becomes clear that bees were on Earth 90 million years earlier. The genus *Homo,* and what's considered its first truly human species *habilis* (handy, tractable), appeared about 2 million years ago, at the onset of the Pleistocene Ice Age. The somewhat more highly-developed *Homo erectus* (upright), who was not only a gatherer but also a hunter with primitive stone tools, followed about 1.5 million years ago. Our species *Homo sapiens* (reasonable, intelligent) didn't appear until 350,000 years ago, at the beginning of the Illinoian-stage glaciation.

A rough overview of the origins and development of several important flora and fauna, including insects, is presented on the next page, in Figure 1.2.

MILLION YEARS AGO

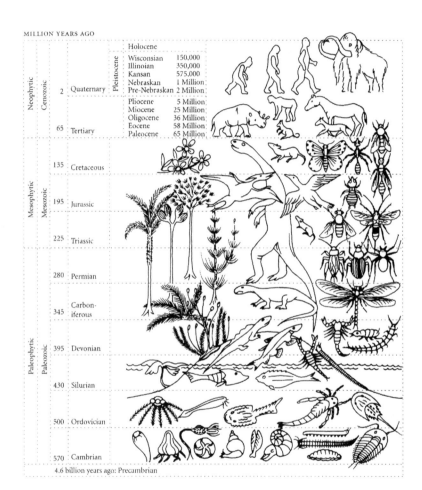

Neophytic	Cenozoic	2	Quaternary	Pleistocene	Holocene
					Wisconsian 150,000
					Illinoian 350,000
					Kansan 575,000
					Nebraskan 1 Million
					Pre-Nebraskan 2 Million
		65	Tertiary		Pliocene 5 Million
					Miocene 25 Million
					Oligocene 36 Million
					Eocene 58 Million
					Paleocene 65 Million
Mesophytic	Mesozoic	135	Cretaceous		
		195	Jurassic		
		225	Triassic		
Paleophytic	Paleozoic	280	Permian		
		345	Carbon-iferous		
		395	Devonian		
		430	Silurian		
		500	Ordovician		
		570	Cambrian		

4.6 billion years ago: Precambrian

FIGURE 1.2

Appearance of the most important life forms through the course of Earth's history, with special attention to insects.

Precambrian: Algae, bacteria, sponges, jellyfish.

Cambrian: Primitive underwater flora (algae), sponges, jellyfish, echinoderms (sea urchins, starfish), primitive snails, brachiopods, cephalopods (nautilus), chitons, jointed worms, onychophores, trilobites.

Ordovician: Graptolites, primitive chordates, jawless ostracoderms, giant see scorpions, horseshoe crabs.

Silurian: Ephemeral land plants (subaerial algae, seaweeds); cephalaspids, jawless coelolepids; millipedes and scorpions leave the water.

Devonian: First land plants (primitive ferns, moss); crossopterygians and amphibians move onto land; land scorpions, millipedes, wingless insects (springtails).

Carboniferous: Tree-like ferns, sigillaria, squamaceous trees, tree-like horsetails, first reptiles, giant insects (primitive fliers).

Permian: First coniferous trees, ginkgo trees, first lizards, hemimetabolous insects (cockroaches, beetles), spiders.

Triassic: Conifers, giant ferns, tree-like horsetails, lizards, small mammals, flies, first Hymenoptera.

Jurassic: Coniferous forests, pterosaurs, birds, termites (?).

Cretaceous: First flowering plants, broadleaf trees, small mammals, digger wasps, ants, bees (upper Cretaceous), butterflies.

Tertiary: Hippopotami and elephant-like giant mammals, large carnivores, developmental stages of horses, squirrel-like primitive primates (Paleocene), transitional period between animals and humans, human-like primates (hominids), and anthropoid apes split (Miocene/Pliocene), social bees (Eocene).

Quaternary: Modern flora and fauna, humans (*Homo habilis*, *Homo erectus*, *Homo sapiens*).

chapter 2

The Bee—An Insect

Before we acquaint ourselves with the "personal" aspects of bees and their community life, we should first get to know them as insects. In order to do that, it's necessary to learn the most important details of insect anatomy, vital processes, and development.

Compared to vertebrates, insects' most distinguishing body feature is their inverse design. Vertebrates possess a bony skeleton, which all their muscles important for movement are attached to. Externally, they have a relatively easily injured skin. In contrast, insect skin is fortified as a sort of armor, which simultaneously functions as a support skeleton for the muscles (exoskeleton). It consists of a very hard nitrogenous material called chitin. The shell-like sections of the exoskeleton are connected with thin moving plates. This structure allows insects a high degree of maneuverability and agility, which is suited to every body size. Among the

smallest insects, for example, is the 2-millimeter long
pharaoh's ant. Walking sticks—which can, in fact, fly—can
be as long as 35 centimeters.

The body of an insect is segmented into three main parts:
the head, the thorax, and the abdomen. The capsule-like
head bears the two many-jointed feelers (antennae) and the
three small, simple eyes (ocelli) that most insects have. It
also holds two compound eyes, which consist of many single
eyes (ommatidia) working together. While the compound
eyes see colors and form, the simple eyes, situated high on
the head in most insects, can only distinguish between light
and dark and are mostly restricted to orientating movement. Insect mouthparts still reflect their original function—chewing (which roaches are still restricted to). They
consist of two simple but powerful jaw-like mandibles for
biting; a pair of jaw-like maxillae (M_1), which hold and
chop the food; and the segmented lower lip (labium, M_2),
which passes on the food. These basic mouthparts differ
greatly among the various insects to suit their diverse life-
styles. Aphids, for example, have a proboscis (maxillae) as
long as their body to pierce the hard surfaces of plants.
Butterflies can roll their up to 20-centimeter-long proboscis
into a tight spiral or unroll it into a straight tube, as needed.
In bees, the mouthparts are a sort of sucking instrument
that can be pulled in or out like a pocketknife. The disap-
pearing tongue is also designed to lap up small amounts of
liquid.

The thorax, which is made of three fixed segments, bears
three pairs of legs and (on winged insects) typically two

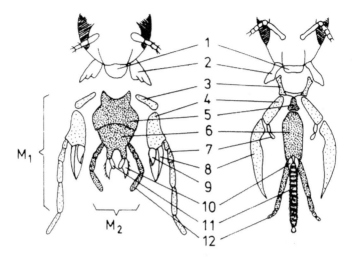

FIGURE 2.1

Mouthparts of the cockroach (left) and the honeybee (right).

1 Upper lip (labrum), 2 mandible, 3 cardo, 4 stipes, 5 lorum (in the honeybee), postmentum (in the cockroach), 6 prementum, 7 maxillary palpus (singular)/palpi (plural), 8 galea, 9 lacinia, 10 paraglossa, 11 glossa, 12 labial palpus (singular)/palpi (plural), M_1 maxilla, M_2 labium.

pairs of wings. The legs serve foremost as locomotion on the ground and as body-cleaning instruments. In addition, they fulfill many different specialized purposes among the various insect groups. For most bees, the legs are especially useful for gathering pollen.

While the head and the nearly completely muscled thorax are relatively fixed body parts, the abdomen, with its multiple telescoping segments, is very flexible. As a rule, the abdomen does not have appendages, assuming that ovipositors and stingers are not defined as such. It contains the great majority of the inner organs. (See Figure 2.2.)

The digestive tube runs through the body from the mouth to the anal opening. The front section is sometimes specialized for certain purposes—a tooth-like chopping mechanism in roaches and grasshoppers, for example. In some insects the abdominal section is inflated to form a sort of crop, called a honey stomach in honeybees. The connecting midgut, or mesenteron, processes the food, which is ready-for-use and absorbed into the blood in the small intestine, the front section of the hindgut. The nephridial tubules (also called renal or Malpighian tubules) are at the entrance to the small intestine. In honeybees, the very rear portion of the digestive tube forms a strong sac capable of enlarging. This excrement sac can retain feces for months during the wintertime.

The vascular system of insects is described as open. It is very simple and consists of a multiple-chambered, tube-like heart in the abdomen, which pumps the blood through a single vessel (aorta) to the thorax. At the end of the vessel,

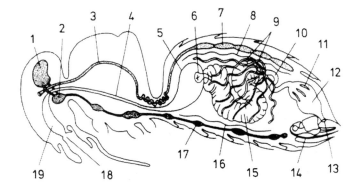

FIGURE 2.2

Cross section of a honeybee with vascular system,
digestive tract, nervous system, and sting.

1 Supraesophageal ganglion (brain), 2 subesophageal
ganglion, 3 aorta, 4 esophagus, 5 honey stomach,
6 proventriculus, 7 heart, 8 dorsal diaphragm,
9 nephridial tubules, 10 anterior intestine, 11 rectal
papillae, 12 rectum, 13 anus, 14 stinging apparatus,
15 midgut, 16 ventral diaphragm, 17 nerve cord,
18 mouth, 19 pharinx.

the blood empties into the body cavity and flows freely over all the organs back toward the abdomen, where it reenters the heart through closable openings. The blood is then recirculated back toward the thorax. Pulsating membranes in the body assist the movement of blood and ensure that it reaches outlying body parts like the wings, the legs, and the tips of the antennae.

Insect blood (hemolymph) is ordinarily colorless and has no red blood corpuscles. Instead, it contains other various blood cells, which, among other things, serve as immune defenses. Like vertebrate blood, insect blood transports nutrients, hormones, and excreta—but no oxygen, which is transported without material carriers.

The respiratory system of insects is also unique but much more complex than the insect vascular system. Its network of air sacs and tubes branch throughout the body to supply necessary oxygen to all the inner organs. The air tubes (tracheae) consist of tightly wound spirals of chitin that connect to external openings (spiracles) positioned on both sides of the thorax and abdomen. The spiracles on the abdomen maintain air circulation by rhythmically closing and opening, using a complex capping mechanism.

The nervous system of insects is arranged like a stepladder. It consists of a brain in the head, from which a pair of parallel nerve strands connected in intervals by ganglia emanate. Numerous nerves branch out from the ganglia into all the body parts. Due to the stepladder nerve pattern on the underside of insects, we speak of a nerve cord.

The reproductive organs of insects are located in the abdomen. The ovaries of the females and the testes of the males develop differently within the various insect groups, sometimes internally and sometimes externally. The position of the genitalia likewise varies, which leads to a diversity of mating behaviors. The abdomen also holds a so-called fat body, or aliphatic compound, which is a reservoir for fat and protein and plays a role in insects' care of their broods and in their overwintering capability. In the social Hymenoptera, the fat body varies in use and size according to the seasons.

Insects possess a large variety of internal and external glands. In addition to the typical secretory glands that discharge either internally or externally through secretory ducts, there are glands that directly release their secretions into the blood. The variety of glands is important for nutrition, finding mates, reproduction, growth and development, and, not least, the social functions of colonies. Many bees have wax glands positioned on different points of the abdomen, depending on the bee family they belong to. These wax glands are used during the building of nests.

When you consider that the 1 million-plus species of insects have colonized every corner of the Earth, from water to land to air, and have adapted to the demands of these many environments, you can imagine how diverse the sensory lives of insects are. Their powers of smell, taste, touch, and hearing encompass all degrees of development from nonexistent to unimaginably highly developed.

Correspondingly, the sense organs of insects are very diverse. For what's relevant to honeybees, we'll return to this subject on page 132.

For all of insects' extraordinary capacities, there's one thing an insect cannot yet do: in its youth, it can't continually grow; the exoskeleton does not allow for this. As a result, insects frequently shed their exoskeleton in a process called molting. The new chitin exterior beneath is initially stretchable, allowing for a spurt of growth. Molting is not only always combined with growth but also often with a change in form called metamorphosis. There are two types of insect development, each of which is shown in Figure 2.3. In the simplest type, a larva (when legless called a grub or maggot; with legs called a caterpillar) hatches out of an egg and assumes the appearance of the adult insect (imago) over the course of numerous single steps. The individual larval stages (instars) closely resemble the adult bodyform. This process is called simple metamorphosis, or hemimetabolism. Roaches, grasshoppers, aphids, true bugs, and termites undergo this type of development. In contrast, in complete metamorphosis, or holometabolism, the larvae do not resemble their adult progenitors at all. They go through an intermediate developmental stage between the larval stage and the adult insect called a pupa. This stage carries out a profound change in the growing insect, whereby all internal and external body structures (e. g., silk glands, pseudopods, gills) disappear and new structures (like wings and stingers) appear. This metamorphosis is triggered by hormones and is most familiar to us in butterflies, which change as pupae

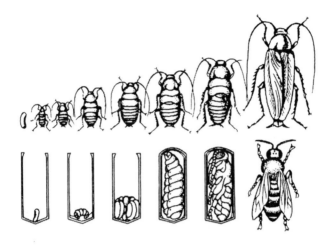

FIGURE 2.3

Top: Hemimetabolic development of the German Roach (*Blatta germanica*) with egg, six larval stages (instars), and adult (imago).

Bottom: Holometabolic development of the honeybee (*Apis mellifera*) with egg, two (of actually four) coiled grub stages, elongated grub stage, pupa, and imago.

from mostly unattractive caterpillars into beautiful winged creatures. Among the colony-building insects, wasps, ants, and bees are all holometabolic. In all three families of Hymenoptera, the larvae spin themselves into pupae, which in ants are very tough and free-standing. Wasp and bee pupae are spun into fine, silk-like cocoons that fit neatly into the cells of a honeycomb. Some insects molt twenty or more times before reaching the pupa stage. Hymenoptera molt a limited number of times before pupating. Honeybees molt five times before becoming pupae. The final, sixth molt (imaginal molt) leads to the fully developed insect.

chapter 3

What Does "Social" Mean in the Animal Kingdom?

The vast majority of insects lives alone and only enters a short-lasting partnership with one or several members of their species during the mating season. This is also true for most bees. But among them, there are also representatives that live in temporary or permanent social groups. We call the most sophisticated forms of these groups *colonies*. The appropriateness of this term in the context of insects is questionable, but it has become so well established in science and practice that it can't be avoided. More neutral substitutes for this not quite flawless term include "animal society" and "animal community." However, many socio-biologists also distinguish between these two terms because they associate "community" with family groups and "society" with unrelated group members. We do not need to adhere to their distinction here.

Bees are fascinating because they include solitary as well as social species and many variations in-between. Before we closely examine this diversity, we should become acquainted with a few other animal societies.

Anonymous Societies

On still days in early summer, you might ask yourself if the thick clouds of mosquitoes gliding up and down over wet ditches and ponds are actually exhibiting social behavior. These swarms result from the biological advantages of cooperative breeding sites and simultaneously represent a violent nuptial spectacle. Location of birth also accounts for the mass appearance of aphids on twigs and leaves, but it does not provide these feeders on plant sap with any mating advantages, since the females produce rapidly successive generations without males through a reproductive process called parthenogenesis. Cooperative breeding exists among a wide variety of animals; it's especially striking among birds. Consider swift colonies in open, sunlit sandbanks and the crowded nesting sites of gulls, rooks, swallows, and cormorants. Weaver finches construct massive communal nests for their colony in addition to building separate nest pouches for their own families. They must, indeed, possess a certain attraction to each other, a tendency toward community.

The development of schools among fish and flocks among birds, as well as packs and herds among mammals, also indicates a social tendency. These societies do not require familiarity between individuals; they can develop out of very mixed groups that have nothing to do with family or species bonds. Thus, wandering herds of hoofed mammals in the steppes often include such diverse species as antelopes, zebras, and gnus. Among birds, jackdaws and crows often flock together, as do various kinds of titmice in winter. Such unions provide these animals with an especially effective defense against predators.

The next step is cooperative rearing of broods. This can be observed among some birds, including penguins, which occasionally incubate eggs from other birds of their species. Rodents sometimes raise their litters together, and mother beasts of prey will feed another mother's offspring next to their own when necessary. We will learn more about the cooperative care of offspring among bees from Chapter 5 on.

As soon as animals form societies and work for each other or exclusively with each other, a mutual hereditary attraction develops and other social impulses frequently take effect. These include imitation, stimulation (mutual encouragement toward departure, flight, feeding, mating, etc.), and synchronization (e. g., movement in the same direction, identical wing beats). Such behavior accounts for the expression "group dynamics."

Animal migrations provide ample examples of group dynamics. Not only mammals (lemmings), birds, fish (spawning), and toads (biotope change) migrate; insects migrate, too. The black fungus gnat larvae (family *Sciaridae*) of North American and European forest soils and the caterpillars of the processionary moths (family *Thaumetopoe-idae*, only found in Europe) offer impressive examples. The latter hold migrations that begin in single file and grow into wide bands that can become meters long. The trigger of this movement is not always nutrient deprivation, even when the migration remedies such deprivation. The tendency to imitate seems especially macabre when such a band of larvae go around in circles. A person can cause this by steering the lead individual toward the tail end of the migrating band. The result is a circling that lasts hours. Another example of insect migration is the notorious movement of locusts, which can cause enormous damage to crops. There have been reports of swarms that included billions of insects covering 100 square kilometers (in North Africa) and others that even reached a length of 100 kilometers (in South America).

Family Associations

The groups of animals described so far are composed of primarily anonymous relationships between individuals. These stand in contrast with social structures founded upon family connections, such as packs of predatory animals that form hunting societies, or the clans of baboons. Nearly all of the higher animals, in fact, display at least one basic element of family organization; namely, that both parents or the mother take care of the offspring, whether they are nursed (mammals), fed (birds), or simply protected, as is the case with catfish and cichlids who keep their eggs in their mouths until the eggs hatch. Frequently, these "mouth-brooders" cultivate a sort of family life for a while, as the parents lead their school of offspring and protect them from danger by letting all the young fish disappear in their mouths.

All family societies develop such protective functions out of absolute necessity. Just as a mother hen leads her chicks away from danger, all mammalian mothers carry their offspring to safety in the face of threats or defend the young ones by risking themselves in fights. Among amphibians, the Surinam toad (*Pipa pipa*) offers a fine example of providing for offspring. The mother carries her tadpoles with her in depressions on her back until they have grown into young toads. Such a sense of family can be observed everywhere, even among insects (arthropods). The common earwig (*Forficula auricularia*), belonging to the primitive order Orthoptera, which includes the grasshopper and the cockroach (in modern classifications the earwigs are separated in a different order, Dermaptera), is a wonderful example of this. Pairs of earwigs mate in September and spend the winter in protective holes under rocks. When the female lays her 40 to 60 eggs in February, she expels the male and busies herself for weeks taking care of her eggs. She pushes the eggs together and sits on top of the pile. If the microclimate changes for the worst, she carries each egg singly in her mouth to a better place. After they hatch, the young earwigs remain in the nest for a period under the watch of their mother. Many kinds of beetles also display refined care for

their offspring. The most pronounced of these are the burying beetles (*Necrophorus vespilloides*). As a pair, they form a
decomposing small mammal or bird into a round lump;
then they bury it and cover it in a hole they have excavated.
The female gnaws a hole into the carcass and lays numerous
eggs in it. When the larvae hatch, the mother feeds them
with her mouth at first, just like bird parents feed their
young. As soon as the larvae bore into the adjacent earth to
pupate, the female dies. Scorpions and spiders also take care
of their offspring. Many species of spiders spin round recep-
tacles for their eggs and later carry their young with them on
their backs.

The Insect Colony

We have purposely not yet touched on the social aspects of
animals in the orders Hymenoptera and Isoptera. These
groups possess social structures that far exceed those we
have gotten to know so far. We will now become acquainted
with the insect colony.

Few creatures in the animal world can compare with the
colony-building insects. One that does is a small, hairless,
vegetarian mammal called the naked mole rat
(*Heterocephalus glaber*), which lives in underground
colonies of 70 to 80 individuals. Among them, only one
especially large female and a few males are sexually active.
The numerous sterile male and female descendants are

dedicated solely to expanding the system of underground tunnels and caring for the reproducing animals and their broods. A beetle (*Australoplatypus incompertus*) has also been discovered that is capable of building colony-like communities. The colony members live in extensive tunnel systems within the heartwood of eucalyptus trees. Many of them are infertile. Apparently, the colony's founding female produces offspring alone. The infertile colony members maintain the tunnels and help with the rearing of the offspring, which include males and new nest founders.

With the exception of such rare examples, we only encounter animal colonies among Hymenoptera and Isoptera. The social accomplishments of the notorious wood-destroying termites of the tropics and subtropics, with their relatively primitive bodies and incomplete metamorphosis, are in no way inferior to those of the most highly developed ant and bee species. No solitary animals exist among today's termite species; the same is true of ants. Only wasps and bees still include numerous solitary forms. In fact, colony builders are the minority among them.

What characterizes an insect colony?

1. It is a family society built by the offspring of one mother—rarely a few mothers. Therefore, we can also define a colony as a *maternal family*. Only termites can be described as having biparental families, since their colonies include a "king's chamber," where the small father, beside the shapeless, egg-laying mother, is also

continually present. In contrast, the males of colony-building Hymenoptera die shortly after mating, just like those who try to mate unsuccessfully.

2. In an insect colony, the young insects take over brood rearing and all other nest work such as bringing in building materials and food. They do not normally take part in the production of offspring, however, because they are typically infertile. Only one insect or a few insects lay the eggs; they are called queens.

3. This division of labor results in the development of more or less apparent bodily differences between the queen and the workers. These are first expressed in body size (e. g., bumblebees and many wasps) but are also apparent in the development of various body characteristics (as in honeybees, ants, and termites). To describe the different body types determined by the division of labor, we speak of different castes (a reference to the Indian caste system), of dimorphism (where two castes are present), and of polymorphism (where many castes are present). Polymorphism exists among various army ant species and among the higher termites, which include half-developed stages and males (so-called soldiers) who also exhibit body characteristics determined by their social roles.

4. Insect colonies survive at least one year (e. g., bumblebee and wasp colonies). Often, they last years, though, as some tropical wasps, stingless bees, and honeybees illustrate. The life expectancy of the queens plays a large role in the duration of a colony's survival. If the

colony rejuvenates itself through swarms or offspring (as stingless bees, honeybees, and ants do), it can actually be considered immortal.

5. As a rule, highly effective and impressive forms of information transference exist among the members of an insect colony. These are partially chemical (glandular excretions, for example) and partially mechanical, such as the informational flights of stingless bees and the dances of honeybees in their hives.

6. Insect colonies are mostly marked by impressively-built living structures: huge, rock-hard termite mounds; wasp nests of artful combs wrapped with fine papier-mâché; the highly precise and symmetrical wax combs of honeybees; and the extensive tunnel systems of some ants that go as deep as 6 meters below the surface of the Earth.

7. The fact that colony-building insects appear in different orders (Isoptera, Hymenoptera) and different families (wasps, bees, and ants) shows that various independent attempts to create colonies existed in the evolutionary history of insects. There was no single leap to the creation of a perfect colony. Instead, many small steps had to be made, and some developments were certainly left behind. Despite all this, the successful insect families have arrived at virtually the same ends of form and function.

Social Classifications

The paths to an insect colony go through many stations. Today, we can distinguish such in-between stages especially easily among bees. In order to make an overview of social development's numerous forms easier, scientists have created a functional ranking of the living genera and the species' various social forms. Since this ranking introduces ideas that are discussed later in the more specialized Chapters 5 through 7, it is only summarized here.

1. Sub-social: Parents (mother) occasionally take care of their brood. In broader terms, this category refers to functional gatherings of related or unrelated insects (locust swarms, aphid colonies, brood colonies, over-wintering groups, etc.).
2. Communal: Insects of the same generation use a shared nest.
3. Quasi-social: Insects of the same generation use a shared nest and work together in brood-rearing.
4. Semi-social: Like quasi-social, with the addition of division of labor for egg-laying and work.
5. Eu-social: Like semi-social, with the addition of over-lapping generations that include offspring who are workers.

Categories 2 through 4 are often combined under the term "para-social."

chapter 4

The Traits of Bees and Their System

Approximately 20,000 bee species have been identified worldwide, most of which live in the tropics. About 5000 bee species have been found to date in the United States. The vast majority of bees remain solitary their whole lives, with the exception of the short mating period. Only a small percentage of bees—not even 15 percent—spend their lives in some form of colony.

Scientists agree that the predecessors of bees were wasps and, more precisely, that today's bees and wasps are both descendants of wasp-like ancestors. Wasps nourish themselves with flesh—especially wasp brood, which only receives flesh. The adults consume mainly sweet plant juices. In contrast, bees are purely herbivorous and nourish themselves completely with pollen and honey derived from plants.

Sometime in their evolutionary history, bees must have switched food sources from flesh to plants. Their ancestors were probably members of the extensive wasp group *Sphecoidea* (digger wasps), which today includes 5000 species worldwide. As already mentioned, these wasps dig holes or tunnels in the ground, where they take various prey (according to their species's preference) that they have paralyzed with a single sting. They lay one egg on the flesh provision, close the tunnel, and thereafter generally do nothing more for the hatched offspring. Many solitary bees nourish their young very similarly, with the exception that they bury pollen, honey, or a mixture of the two instead of an insect. This has two advantages: collecting plant material is not as dangerous as killing prey, and plant material keeps longer than flesh.

In order to gather pollen, bees need certain anatomical adaptations. While the surface of a wasp's body is smooth and nearly hairless, almost all bees have a thick covering of branched, feathery hairs. These hairs are used to collect pollen. Only the members of the primitive genus *Hylaeus*— the masked bees—which are nearly naked and possess wasp-like mouthparts, are not able to gather pollen this way. They must swallow the pollen in order to bring it home in their crop. The more highly developed carpenter bees (*Ceratina* and *Xylocopa*) also store pollen in their crops, though they do not need to since they are well covered with hair. All other bees carry pollen on the outside of their bodies. Often, their hind legs are designed for this. The top of Figure 4.1 shows the middle leg of a honeybee, which is a prototypical, unspecialized insect leg used mainly for locomotion. The drawings below picture the body parts several

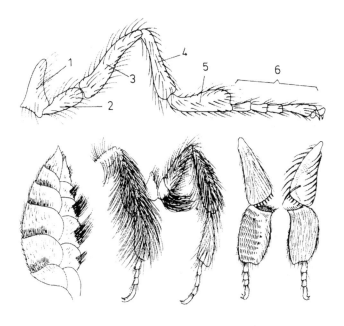

FIGURE 4.1

Top: A honeybee's middle leg is a prototypical, unspecialized insect leg. 1 Coxa, 2 trochanter, 3 femur, 4 tibia, 5 basitarsus, 6 mediotarsus and distitarsus.

Bottom: Pollen-collecting structures of *Apis mellifera*, the western honeybee. From left to right: gastral (or metasomal) scopa (*Megachile versicolor*), tibial scopa (*Dasypoda plumipes*), scopa with a floccus (tuft of hairs) on the trochanter, and other scopal hairs on femur and tibia (*Andrena clarcella*).

different bees use for transporting pollen. They are so different among their respective species and yet so typical that they—and the pollen-collecting habits they represent—have been used as the basis for systematizing all bees.

In addition to bees that collect pollen using their legs, there are bees that carry pollen on the underside of their abdomen using a layer of thick, sometimes brush-like hairs. These bees include the carder bees (*Anthidium*), resin bees (*Heriades*), mason bees (*Osmia*), leaf-cutting bees (*Megachilidae*), mason bees of the walls (*Chalicodoma*), and megachilid carpenter bees (*Lithurgus*). When these bees rub against the stamens of a flower, pollen sticks to their abdomen. They also collect pollen by moving their abdomen up and down in a flower, which causes pollen to brush off the stamens more completely.

Most other bees carry pollen on their hind legs. Among them, we discern two main groups according to the location, thickness, distribution, or other particular characteristics of the bees' hair—those who collect pollen on their femurs (e. g., andrenid bees [*Andrenidae*]), and those who collect pollen on their tibiae (such as miner bees [*Panurgus* and *Anthophorinae*]). Many transitional variations exist between these two groups, just as among the bees that collect pollen on their abdomens. Some bees carry the pollen home dry, while others moisten it slightly with honey from their honey sac.

One more group—the pollen-basket collectors—differ quite significantly from the rest. This group includes bumblebees, stingless bees, and honeybees, all of which

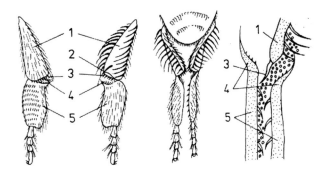

FIGURE 4.2

Pollen-collecting devices on the hind legs of the honeybee.

From left to right: The view from inside, from outside, from behind, and a cross section of the tarsi and tibia that shows the action of the pollen-collecting devices.

1 Tibia with pollen basket (on the outside), 2 single corbicular bristle, 3 pollen comb, 4 pollen press, 5 tarsi with pollen brush (on the inside).

manage to store pollen in solid bunches on the tibiae of their hind legs. Their pollen-collecting tools consist of the brush—multiple rows of bristles on the inside of the enlarged and flattened hind tarsi—the comb on the bottom edge of the hind tibiae, and the pollen press—a projection on the upper edge of the tarsi, opposite the comb (see Figure 4.2). The pollen is collected in a marvelous process that proceeds as follows: Through skillful cleaning of all its legs, the bee gathers pollen it has accumulated from landing in a flower onto the brushes of its tarsi. While the bee is flying, the comb of one leg combs out the pollen from the brush of the other leg, and vice versa. When the tarsus and tibia rub against each other, the pollen press pushes against the upper edge of the tarsus, which moves the pollen out of the comb up onto the outside of the tibia. Here, long, arched bristles create a sort of basket, which collects the pollen. This process happens so fast that nothing can be seen with the naked eye except for the hind legs rubbing against each other.

Not all solitary bees dig nests in the ground for their broods like their wasp ancestors did. Most species of the following families, however, do: masked and plasterer bees (*Colletidae*), andrenid bees (*Andrenidae*), melittid bees (*Melittidae*), halictid bees (*Halictidae*), and some miner bees (*Anthophorinae*). Other solitary bees use found holes in plant stalks, tree stumps, walls, and crevices, which they enlarge, if necessary. These include most *Megachilidae* (leaf-cutting bees and mason bees) and various *Anthophorinae*. Within these holes, they erect partitions between the brood cells out of resin, wood shards, or mud, using saliva to bind the materials together. Several *Megachilidae* of the genus *Osmia* prefer to nest in empty snail shells. Others build brood cells out of resin or mud mixed with sand and small stones and attach them to stone walls or cliffs. As a rule, the cells are carefully lined, regardless of whether they are constructed in the ground or elsewhere. For lining, masked bees, plasterer bees, andrenid bees, and miner bees utilize secretions from salivary glands and/or abdominal glands. Carder bees wallpaper their nests with gnawed plant hairs; leaf-cutting bees use pieces of leaves. We will discuss this in more detail in the first part of Chapter 5.

Wild bees' relationships to their food sources are just as diverse as their nesting habits. Some species visit the most varied plants in a quite arbitrary fashion. Others are so choosy that they depend on a single plant or just very few plants for their food. The common names of such specialists often reflect their preferred food plants; e. g., winterheath-plasterer bees, comfrey-mining andrenid bees, bryony-mining andrenid bees, red-bartsia mellitid bees, purple-loosestrife mellitid bees, bellflowers mason bees, blueweed mason bees, buttercups resin bees, and so on. The length of a bee's proboscis, which is different from species to species, also determines the flowers a bee chooses to visit. Species such as *Hylaeus* (masked bees) and *Colletes* (plasterer bees), with their 2- to 3-millimeter-long probosces, are especially suited to flowers with easily-accessible nectar (e. g., butter-cups, crucifers, umbels, and composites). Species with long probosces (7 to 9 millimeters), such as carder bees (*Anthid-ium*), large carpenter bees (*Xylocopa*), and miner bees

(*Anthophorinae*), also visit deep-blossomed flowers like peas, labiates, and figworts.

The nutritional specialization of many wild bees should discredit some conservationists' claim that honeybees, with their large colonies, compete with solitary bees for food and could drive these wild bees away from their home ranges. This is very improbable. Due to their highly developed forms of communication, honeybees work very economically. They primarily search for food sources that will be worthwhile for a whole colony to find, such as agricultural plants, orchards, and conifer plantations. Thus, they generally overlook single or small stands of blooming plants, which provide the main sources of pollen for wild bees. According to all existing observations, honeybees and wild bees live in peaceful coexistence, not in competition.

Since wild bees have so many surprisingly different nesting and pollen-collecting habits, it seems logical that these characteristics would play a meaningful role in the taxonomic classification of bees. They are, in fact, helpful, but anatomical characteristics, such as head shape, mouthparts, length of proboscis, male reproductive parts, and pollen-collecting devices, play a larger role. The body size of the adult insect is also important. There are miniature bees that only become 4 to 5 millimeters long, like some masked bees (*Hylaeus* species) and the tiny sweat bee (*Lasioglossum microlepoides* in North America, or *Lasioglossum pauxillum* in Europe). Others grow to almost 3 centimeters long, such as the large carpenter bees of the genus *Xylocopa*.

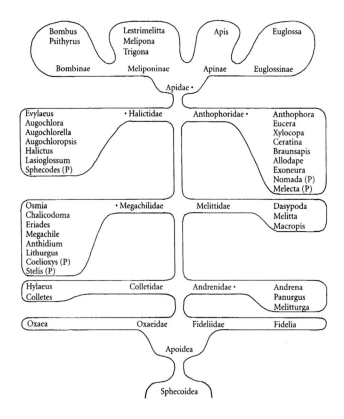

FIGURE 4.3

Tree diagram of the nine bee families and selected genera mentioned in the text. The families marked with • include various forms of social species. (P) denotes the occurrence of parasitic, or cuckoo, species within the genus.

Given the enormous number of bee species, it is no wonder that their systematic classification into related groups presents all kinds of difficulties. Scientists do not always agree about their classification. It would demand too much of us to go into taxonomic fine points here. We need an overview, however, and Figure 4.3 provides us with one in the form of a tree diagram (but this should not be viewed as a family tree of bees' evolutionary history). According to the broad consensus among taxonomists, bees can be divided into nine families. We can disregard two of these families, the *Oxaeidae* and the *Fideliidae*, since they include only a very few exotic species that are unimportant for our purposes. We will encounter many genera from the rest of the families in the following specialized section of this book, and we will focus only on these because the incredible diversity of genera names quickly becomes overwhelming. The numerous, often contentious subfamilies and tribes that taxonomists have ordered into subdivisions will also remain unnamed (except for the *Apinae*).

Among the 20,000 bee species on Earth, the highly social forms include the fewest species. The bumblebees (*Bombinae*) number 200; the stingless bees (*Meliponinae*) just over 300; and the highly social honeybees (*Apinae*) include just 9 species distinguished from each other so far. The exclusively tropical orchid bees (*Euglossinae*)—with their large, beautiful, metallic green, blue, red, and purple representatives that mainly pollinate orchids—belonged to the *Apinae* in earlier times but stopped halfway in the development of colonies.

chapter 5

Solitary Bees and Social Development

Given the enormous number of solitary bees and bees on the way to becoming social, we must content ourselves with describing several selected, especially informative and impressive examples. Whenever possible, we will examine bees that occur in North America.

Solitary Bees

We will begin with bees that belong to the primitive family *Colletidae*—the masked and plasterer bees. Masked bees (*Hylaeus*, or *Prosopis*, species) have many wasp-like features. They are nearly hairless, possess a short tongue unsuited to sucking, and are also small: only 4 to 8 millimeters long. At least 500 species of masked bees exist, ranging from hot to cold climates and all altitudes over almost the entire Earth.

Roughly 50 species are found in the United States. Their English name comes from the characteristic white or yellow mark that most of them have on their faces. It is especially distinct on the males. Otherwise, they are primarily black. For their nests, masked bees prefer blackberry brambles, reeds, or similar hollow stems but will also choose porous clay walls, cracks in wooden posts, and abandoned beetle tunnels. The females always work alone, penetrating cracks or bitten holes in a hollow stem, removing a portion of the plant medulla, then lining a cell with a thin, transparent, watertight secretion. This material comes from the Dufour's gland on the abdomen, and saliva is added to it as the tongue spreads it. Next, the female brings nectar and pollen from her crop into the end of the cell and stores the mixture there. She then lays one egg on the ball of food and closes the brood cell with a horizontal wall made from hardened glandular mucus. Just before that, however, she has prepared the next cell and a row of others—up to a dozen or more (see the top of Figure 5.1). But the female does not always utilize the available space efficiently. Often, she leaves bits of plant medulla between the cells, and, generally, a section between the last cell and the entrance to the nest is left open. When the female is finished with the cells, she seals the main entrance to the horizontal nest with an especially solid glandular secretion. Then she abandons the nest site, never to pay attention to her brood again. The hatched larvae nourish themselves with the provided balls of food and spin a cocoon around themselves for the winter. They don't pupate until the following May, when after a month or so—

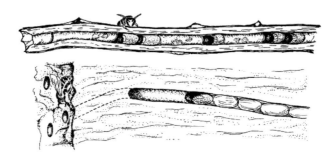

FIGURE 5.1

Top: A masked bee's horizontal nest in a blackberry bramble.

Bottom: Nest tunnel of a plasterer bee (*Colletes cunicularius*) in a sandbank.

depending on the weather—they emerge from their cocoons as young masked bees and work their way out of the nest. The males appear out of the front cells somewhat earlier than the females. This phenomenon is called proterandry. After mating, the males die, while the females live on to make the necessary preparations for their brood. Given good summers, two successive generations of masked bees should be possible within the same year.

Like masked bees, plasterer bees (*Colletinae*) line their rows of cells with a mixture of saliva and glandular secretion that hardens into a cellophane-like membrane. There are not as many species of plasterer bees as there are species of masked bees. In the United States, 120 species can be found. Thick, yellowish-gray hairs cover the head and thorax of plasterer bees, while their abdomens are banded gray. Their hairy hind legs are well suited to collecting pollen. Rather than using existing tunnels for their nests, plasterer bees dig their nest tubes in the sides of earthen walls, regardless of whether they are light sand or heavy clay. Most often, the round, roughly 5-millimeter-wide tunnel initially slants up before it turns as many as 10 centimeters downward. Each female plasterer bee fashions roughly ten brood cells separated by plaster-like walls and leaves a honey-pollen clump and one egg in every cell (see the bottom of Figure 5.1). As with masked bees, the young do not pupate until the next year. Emergence and activity times are different among the various species and begin either in early spring or first in summer. Often hundreds or even thousands of plasterer

bees nest in small spaces without disturbing each others'
tunnels. They apparently also reuse old abandoned nests.

The worldwide, very species-rich andrenid bee family
(*Andrenidae*) is represented by a little over 1300 species in
the United States. Some andrenid bees look almost like
honeybees. Others are larger; yet others are only 4 to 5 millimeters long. The females are typical of bees that gather
pollen on their femurs, with a conspicuous, thick curl of
hair (floccus) on the trochanter, which is the small leg
segment found between the coxa and the femur.

All andrenid species nest in the ground and dig tunnels in
sand, loess, or clay that go from a few centimeters to up to
half a meter deep. Bottle-like cells branch off from the main
tunnel (see the left drawing in Figure 5.2). The walls are
permeated and solidified with a secretion from the Dufour's
gland. After the female has provided every cell with a rich
pollen-honey cake and an egg, she closes the brood chamber
with a cover made of saliva-filled soil. Some species gather a
small pile of soil over the entrance to the tunnel, which they
repeatedly open and close while they collect material for the
nest (see the right drawing in Figure 5.2). The young remain
in their brood cells for the winter as a rule but develop into
fully-grown adults within the same year. The delicate males
of *Andrena vaga* (the European willow miner bee) are the
first harbingers of spring, often appearing out of the earth
before March. They fly back and forth in thick droves over
the ground, looking for the females that appear somewhat
later. Andrenid bees like to nest in parks and gardens, where

hundreds of their tunnels often lie close together. Although each female works strictly for herself, many females of some communal species, such as a miner bee (*Andrena jacobi*) and two panurgines (*Panurgus calcaratus* and *Melitturga claviceps*), have been observed using the same nest entrance. In southern regions, some andrenids are said to raise two generations within one year.

The family of leaf-cutting bees, mason bees of the walls, and mason bees—*Megachilidae*, named after the globally dispersed genus *Megachile*—is extraordinarily diverse and species-rich. All megachilids collect pollen on their abdomen. They can be easily identified by their abdominal brush composed of stiff, backward-pointed bristles whose coloring clearly contrasts with the other abdominal hairs.

The smaller group of leaf-cutting bees (*Megachile*) is primarily at home in subtropical and tropical regions. Over 140 species of leaf-cutting bees can be found in the United States, and 80 more in Mexico—all between 1 and 2 centimeters long and characterized by a dark coloring. The abdomen of a leaf-cutting bee often appears somewhat flat and extraordinarily flexible, even bending over the body to the head. It carries an occasionally fox-red, sometimes white or black brush. Female leaf-cutting bees expand cracks in dry wood or dry walls for nests, dig tunnels in sandy ground, or use hollow plants stems. They carefully paper the walls of their brood cells with pieces of leaves from lilac, rose, raspberry, or other woody plants, but seldom with leaves from leafy plants.

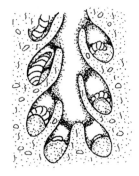 

FIGURE 5.2

The miner bee (*Andrena vaga*) builds a 20-
to 40-centimeter-long vertical tunnel in the
ground, which branches into multiple
bottle-like cells at its end (left). At the nest
entrance is a tower-like pile of soil, many
centimeters high, with a side entrance hole.

With her sharp mandibles, a female leaf-cutting bee saws oval pieces from leaves, rolls them together, and typically carries them tightly under her body to her nest site. Here, she begins wallpapering at the end of the brood tunnel while forming a thimble-like container out of the leaf pieces. Thereafter, she puts down a base of pollen and honey in the first cell, lays an egg on it, and closes the cell by constructing a wall made from round pieces of leaves. She then repeats this process until a half dozen or more cells are full, and finally closes the tunnel with a many-layered deposit of round leaf pieces. *Megachile versicolor* prefers to use rose and blackthorn leaves. The top of Figure 5.3 shows this bee working. The young *Megachile* positioned in their cells do not complete their development until the following spring. In the front cells, the males hatch first, as is customary among many bees.

Instead of leaves, the wasp-like, yellow and nearly hairless megachilid carder bee (*Anthidium*) uses stem fibers from leafy plants such as mullein, hedge nettle, and red dead nettle to line their brood cells. Like leaf-cutting bees, they too build their nests in holes in the ground, cracks in walls and cliffs, and hollow plant parts. Carder bees, such as *Anthidium punctatum* with its small abdominal dots, carry small balls of plant fibers into their nests. With the help of saliva, they form these balls into a kind of hollow cotton wad, which serves as a brood cell (see Figure 5.3, lower left). To provision the cell, a female carder bee fills it first with honey, then turns herself around and spreads pollen on the honey. Between flights to gather the food, she repeatedly

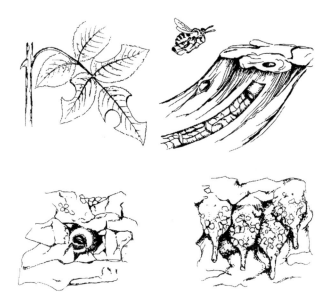

FIGURE 5.3

Top: The leaf-cutting bee (*Megachile versicolor*) cuts oval and round pieces from rose leaves and carries them rolled-together under her body, in typical fashion, to her linear nest tunneled into wood. The rolls of leaves serve to line the cells.

Bottom left: A female carder bee (*Anthidium punctatum*) disappears in the last cell of her linear nest, located in a crack in a wall.

Bottom right: The cells of another carder bee (*Anthidium stigatum*) are, well camouflaged, attached to a stone.

pushes the cottony cover of the cell together to keep out intruders. After the egg has been laid and the cell has been closed, this textile worker of a bee constructs another cell above or next to the previous one. She continues this process until she reaches the entrance of the nest, which she then barricades with small stones or pieces of wood.

Bees that use tree resin to build their nests also belong to the carder bees' group. One of these is *Anthidium strigatum*, a small carder bee. This 6- to 7-millimeter-long bee with a white abdominal brush looks for her nest site on tree trunks or cliff walls, where she attaches small groups of cells in holes and also sometimes on the surface without protection. She collects resin with her mandibles and carries clumps of it to the nest site under her head between her front feet. From the mostly tiny particles of resin, she forms jar-like cells with openings on their bottoms. After the female carder bee has provisioned a cell with food and laid an egg on the provisions, she closes it to a small, tapered point, leaving a tiny air hole at the end. The young bees spend the winter as larvae in the cells and pupate in spring. They leave through openings they gnaw out in the bottoms of their cells, appearing relatively late (not until late June) in the open air. About 30 carder bee species live in the United States.

Mason bees (*Osmia*) are the most diverse genus of *Megachilidae* in their choices of nest sites and building materials. More than 130 species of mason bees live in the United States. Their thoraxes are mostly black, but also metallic green, yellow, or copper-colored, and their stocky, almost cylinder-like bodies are very conspicuous. They build their nests in diverse locations, according to their species' preference: in the ground, in wooden beams, hollow stems, empty snail shells, or attached to stone walls. The cells are arranged either next to each other, behind each other, or above each other. For building materials, mason bees use chewed-up leaves and flower petals or make a mortar out of clay, sand, and saliva. The red mason bee (*Osmia rufa*) is a common bee in central Europe; it is roughly 12 millimeters long and covered with thick, yellowish-red hairs. Female red mason bees search for long nest cavities in walls, beams, or reeds (see the top of Figure 5.4), where they build linear or irregularly grouped nests of up to 20 cells.

Red mason bees separate the cells with clay mortar, which they carry to their nest in small balls. Before a female abandons her nest, she carefully closes the entrance with an especially thick layer of mortar. The young red mason bees develop fully well before the outbreak of winter, but they remain in their protective cocoons until the time for them to fly arrives in the following year. To help the males emerge before the females, the mother only lays male eggs in the cells closest to the exit. They must break through the thick wall of mortar to glimpse daylight.

Other *Osmia* bees work similarly to *Osmia rufa* but separate their cells with resin (like *Heriades truncorum*, the resin bee) or a sort of cement made out of chewed-up plant parts (*Osmia caerulescens*). The mason bee *Osmia mustelina-emarginata* builds egg-shaped cells out of plant cement in protective holes in walls. These cells are, like the jar-shaped cells of carder bees, left open at the bottom while the female gathers food provisions and are then closed with cement. A pile-like nest of brood cells arises, which ultimately creates a protective wall of empty cells for the site's continued use as a brood area. Often, mason bees build their small settlements on the outside of cliffs, and communal mason bees have been observed working together on these sites.

The poppy mason bees (*Osmia papaveris*) dig one bottle-like cell for each of their offspring in sandy soil. Like leaf-cutting bees, poppy mason bee females wallpaper each cell with pieces of leaves, though poppy mason bees use primarily poppy petals. After she has folded the petals against the top of the cell, each female covers the tiny nest with a layer of sand, which she carefully smoothes. Nearby, she repeats this process, preparing multiple single cells. The larvae already spin their cocoons to pupate 14 days later, and another 14 days after that they are ready to leave the nest. After mating, the males die and the females crawl into holes in the ground or cracks in trees, where they spend the winter. During especially long and warm summers, though, they can already begin producing offspring within the year they are born.

Osmia bicolor is a very specialized nest builder. This beautiful black, 9- to 11-millimeter-long bee with bright-red abdominal hairs searches for an empty snail shell for each of its eggs. The female then stores the typical pollen-honey provision deep inside the shell, lays her egg on it, and closes the brood chamber with a sideways wall made from chewed-up leaves. She fills the remaining space in the shell with pebbles and builds another protective wall made of

FIGURE 5.4

Top: Linear nest of *Osmia rufa* (the red mason bee).

Bottom left: Snail shell nest of the mason bee *Osmia bicolor* with a camouflaging roof of stalks and needles; next to it, the cell and cell walls within the snail shell.

Bottom right: Small collection of a mason bee of the walls (*Chalicodoma muraria*): walled, jug-shaped cells partially covered with fine plaster.

hardened leaf-puree in the entrance to the shell (see Figure
5.4, lower left). Thereafter, she frequently turns the shell,
keeping the opening on the bottom, which is tedious, difficult work.

If there is a nearby crevice in the ground, the female digs
the shell into the crevice and moves the lump of earth out of
the way. Finally, she makes numerous flights carrying dry
stalks, pine needles, and thin twigs to the nest, which she
uses to cover it like a tent so it cannot be seen. She also
weaves tiny pieces of moss and grass into the tent and sticks
it all together with saliva so the tent cannot be blown away
by the wind. (A person could still easily pick up the whole
covering, though.)

Of course, this whole nesting process takes time. If bad
weather doesn't slow it down, at least two days are necessary.
And all of that for just one offspring! One female does not
accomplish more than six or seven nests in her entire life,
assuming she can even find the needed snail shells. Other
"helicophile" mason bees (those that nest in snail shells)
work somewhat more economically by building not just one
cell, but many behind each other within one shell. For this,
they also select appropriately larger shells. Interestingly,
there are only two snail-nesting species in the United States:
Osmia conjuncta and one species of the exclusively American genus *Ashmeadiella*.

Instead of using already finished nest sites, the megachilid *Chalicodoma muraria* uses self-made nests. This
female mason bee of the walls combines grains of sand with
water and saliva to make a kind of coarse cement, which she

fashions into jug-like containers open at the top and attached to stone walls and cliff faces in narrow rows. As soon as each container is provisioned with food and an egg, the female closes it with cement. In this fashion, a small collection of cells develops, typically a dozen or fewer. When the female is finished building, she works as a stucco-plasterer, smoothing the whole group of cells with a fine plaster (see the bottom right of Figure 5.4). The mortar becomes exceptionally solid, which makes you wonder how the young mason bees of the walls can gnaw out of these prisons. It appears that they not infrequently require two years in the cells before they can free themselves.

At any rate, mason bees of the walls apparently repair and reuse old nests. *Chalicodoma sicula* bees have been known to completely pave over the hollow grooves in Greek temples. Another mason bee, *Osmia caementaria*, erects nests similar to those built by mason bees of the walls, comprised of five to ten cells covered with a fine plaster. *Osmia caementaria* bees make their cells only half as large and use a coarser building material. Sometimes you can see larger collections of up to 30 cells, which numerous females work together on. As a rule, these nests are used multiple times.

More than 1000 of the widely dispersed digger, cuckoo, and carpenter bees of the family *Anthophoridae* can be found in the United States. The genera can be very diverse in appearance and behavior. Several of them comprise the subfamily *Nomadinae*, whose representatives resemble wasps with their sparse hair covering, yellow or bright coloring, and short probosces. These "cuckoo bees" are also an exception among anthophorids because they live at the expense of their relatives: they are parasitic. The vast majority of other anthophorids are large, with stocky bodies and thick coverings of hair that make them easily to be confused with bumblebees. Like bumblebees, they are equipped with an especially long proboscis, which allows them to feed on flowers with deep-lying nectar (peas, labiates, and figwort). They collect pollen on their legs and dig their nest holes in clay soil or, not infrequently, in clay-plastered walls. The mostly short, sometimes forked nest tunnels end either in multiple back-to-back or grape-shaped branches of egg-like cells. Anthophorids provide their cells with smooth inner walls by covering them with a white, parchment-like mixture apparently made of saliva and excretions from abdominal glands. The cells are separated from the main entrance and from each other with clay walls. Often, enormous brood colonies are found together. The digger bee *Anthophora parietina* (also *A. plagiata*) mixes the soil excavated from digging her nest with water to build a tubular structure in front of the nest entrance (Figure 5.5, left). The clay walls where these bees nest can be decorated with many hundreds of these tunnels slit at the end.

The carpenter bees (*Xylocopinae*) are also customarily considered a subfamily of anthophorids. They resemble bumblebees and are very large. However, they don't nest in mineral material but in dead, usually still somewhat solid wood. In her typically self-chewed tunnel, the female carpenter bee arranges her cells behind each other and divides them with walls made of wood shards stuck together with

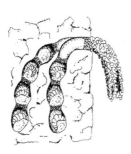

FIGURE 5.5

Left: *Anthophora parietina* lengthens her nest entrance with a tunnel-shaped structure that opens in a slit at the end.

Right: The melittid bee *Dasypoda hirtipes* forms food provisions from the pollen and honey on her long brushes and stores them on three small pedestals to prevent them from becoming damp with soil moisture.

(*Friese*, Die Europäischen Bienen, *Berlin, Germany: Walter de Gryuter, 1923; after an example from A. Giltsch.*)

saliva. Carpenter bees live only in warm regions, where they belong to a subfamily (*Ceratininae*) that shows the beginnings of social behavior. In sunny locations in central Europe, the bumblebee-like, blue-black *Xylocopa violacea* can be observed, which is up to 28 millimeters long. Several similar species of *Xylocopa* can be found in the United States.

With only a few genera and species, the melittids (*Melittidae*) are a small, somewhat incongruous family of bees, which all collect pollen on their legs. They nest in sandy soil, wherein they often dig tunnels of substantial depth (up to 60 centimeters). At the end of these tunnels, they fashion round cells. Several bees of the genus *Melitta* prefer very specific plants for nectar and pollen, including bellflower (*Campanula*) and red bartsia (*Odontites*). The *Dasypodinae* also belong to the melittid family. They are characterized by an especially long pollen-collecting brush on the hind legs, which enables them to carry an amount of pollen equal to half their own weight. There are over 20 species of *Dasypodinae* in the United States. These bees don't simply store provisions for their offspring on the floor of the brood cells, but instead put honey-pollen clumps on three small pedestals that prevent the food from becoming damp (Figure 5.5, right). The females of *Macropis labiata* accomplish similar feats of pollen transport, and, in contrast to nearly all wild bees, they prefer wet places on the banks of ditches and streams for their nesting sites.

On the Way to a Colony

All of the wild bees discussed so far are solitary. After mating, female solitary bees take over all the work necessary for the perpetuation of their species, though this only includes building a nest and providing the brood cells with provisions. Once all the cells have food and an egg, the mother closes the nest and does not concern herself anymore with her offspring. Occasionally, nest sites become substantial collections of cells. Although hundreds, even thousands, of solitary bees work only a few centimeters away from each other at these sites, they do not, as a rule, cultivate any neighborly contacts. We cannot conclude that these bees exhibit social behavior just because they share a nest site. They simply find themselves together at especially good nest building sites or places with favorable soil composition. Perhaps they also just remember where they were born and return there after mating.

Despite all "personal independence" in the brood colony, there are collective actions. Heinrich Friese (1860–1948), the father of wild-bee research, reported that he was suddenly attacked by a swarm of loudly buzzing miner bees (*Andrena vaga*) when he swung his net in the flight path of their 300-nest colony. The bees attacked him so violently that their impact on his body made them fall to the ground. Friese was attacked just as aggressively by common digger bees (*Anthophora acervorum*) that were nesting by the thousands in the dirt wall of a barn. (They had worked it so heavily that it ended up looking like it was riddled with shotgun pellets.) Such things happen only among large groups of solitary bees. Single-nesting females do not show that sort of aggression. Friese's colleague von Buttel-Reepen concluded that it is the quantity of bees that makes courage. He intruduced the theory of "social behavior under exceptional circumstances."

Another first step toward social behavior can be discerned among some digger bees (*Anthophoridae*). The previously mentioned carpenter bees (*Xylocopinae*), with their 200 to 300 mainly tropical species, hatch already in the year of their birth. However, they don't mate then; instead, both males and females search for a place to overwinter. Not infrequently, they use the abandoned nest tunnels of other *Anthophora* species, where they often find themselves gathered together. Dwarf carpenter bees (*Ceratina*), whose many species include more than 20 in the United States, are a subfamily of *Xylocopinae* and exhibit similar behavior. These bees with thick, club-like ends on their antennae prefer to nest in *Rubus* stems, where they divide their cells with walls made of pulp watered down by saliva. The hatched offspring spend the winter in groups of as many as dozens, which widen passages in hollow-stemmed plants they have sought out. Some researchers interpret their sparse hair covering and spiked tibiae as secondary characteristics suited to the crowded nature of their winter quarters. Overwintering groups also exist among several lower forms of *Halictidae* (halictid bees), which will be discussed shortly. On occasion, these groups actually dig their overwintering tunnels in the ground together.

The purpose of such gatherings for winter appears to be protection from the threatening effects of the weather. When other bees, primarily males, gather in sleeping clusters on exposed plant stems in springtime, we search for similar apparent reasons. While they scatter in all directions during the daytime, these bees repeatedly select a certain plant stem to gather on as soon as twilight falls or bad weather arrives. Holding on tight with their mandibles or legs, they line up behind each other and spend the night where they receive neither food nor protection from cold, rain, and wind. Each bee would be better protected inside any flower. Despite this, such sleeping habits have been observed in nearly every bee family. But the reason why male bees gather to sleep in the same place remains a mystery.

If you have a difficult time acknowledging overwintering and sleeping groups as socially motivated and prefer to think they are coincidental gatherings, you might more readily accept that females, which use the same entrance to their nest sites, exhibit a social inclination. Scientifically, this type of social behavior must be classified as "communal." Many diverse species of digging bees have been observed using a shared nest entrance: the andrenids *Panurgus calcaratus* and *Andrena jacobi*; the anthophorid *Eucera longicornis*; the halictid *Halictus longulus*; and others. These females use the same nest entrance but build their own cells and provide them with food and their own eggs. When the nest entrance occasionally becomes crowded, these bees show mutual consideration. They even assign one bee to guard the nest entrance and close it with her head. *Halictus longulus*, which creates nesting communities of 20 to 30 bees, exhibits such guarding behavior. When members of the nest community want to enter or exit, the guard readily retreats to allow room. If you catch a guard, a new one immediately replaces him. If you don't catch the guard within the first few attempts, he turns around and points his stinger outward. The entrance is walled shut if you continue to disturb the guard.

Beyond these modest beginnings, digging species, especially, show even clearer steps toward social development. The members of the large sweat bee family (*Halictidae*) clearly demonstrate stages in social development from solitary to various highly social forms. The halictids are probably the most species-rich family of bees in the world.

They were once all classified as part of the genus *Halictus*, but modern taxonomists now distinguish 30 genera among them (*Halictus, Lasioglossum, Evylaeus, Dialictus, Augochlora, Augochlorella, Augochloropsis, Sphecodes*, etc.). More than 500 species live in the United States, which are not always easy to distinguish from each other. These primarily black bees are 5 to 15 millimeters long, depending on their species, and gather pollen with their legs. Their hind legs are thickly covered with hair. Their English name, "sweat bees," comes from the habit of one of the subfamilies (*Halictinae*) of licking off sweat from people's faces.

Halictids nest in sandy or clay layers of soil, where they dig tunnels up to 30 centimeters deep. Many add their brood cells individually or in grape-like bunches on the sides of a tunnel or tunnel system (see the left and center drawings in Figure 5.6). If their nest entrance is at ground level, halictids occasionally build a pile of saliva-hardened diggings many centimeters high around it. Several skillful species join their cells together in a sort of comb that is attached to the surrounding cavity with bridges of soil. By permeating each part of the nest with water-repelling secretions from the Dufour's and saliva glands, the entire structure of cells gains strength and firmness. A divided entrance and exit to and from the nest provides optimal ventilation (see Figure 5.6 right).

The young bees leave the nest in the same year they are born, and the mated females overwinter in old tunnel systems in the ground. The lower species live alone like true solitary bees, nevertheless generally in considerable colonies that use communal entrances and guards. Higher species create quasi-social communities in which the members work together to build cells and provision them. Relationships are even further developed among the South American species *Augochloropsis sparsilis*, which prefer to nest not in the ground but in rotten wood. In these bees' nests, only some females lay eggs. The others, which possibly returned to their nest after unsuccessfully trying to mate, take over all the rest of the work such as building cells, cleaning, bringing in food, and guarding the nest entrance. This division of labor occurs not only among bee siblings but also among random groups of bees nesting together. When considering the origin of bee colonies, this is a fundamental point to recognize. This kind of community life and

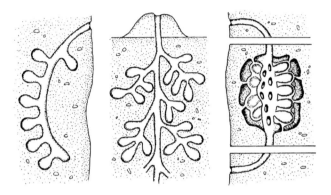

FIGURE 5.6

Ground nests of sweat bees.

Left: Basic type with main tunnel and adjacent bottle-shaped cells (*Halictus sexcinctus*).

Center: Tunnel system of cooperative-building siblings (*Evylaeus malachurus*).

Right: The cavity nest of *Halictus quadricinctus*.

activity among unrelated animals is a dead-end street. We learned the term "semi-social" to describe this sort of cooperative work.

The potential for an insect colony to develop is much higher when all of the female workers descend from the same mother. In central European sand pits and path embankments, *Halictus sexcinctus* can be found. These bees build a simple nest with single cells adjacent to a main tunnel (see Figure 5.6 left), wherein the nest founder still lives after her young have hatched. However, she is no longer of use to them. The female offspring work for themselves, though many generations may use the same nest. European *Halictus quadricinctus* behave similarly in their cavity nest. The nest mother builds and provisions her cells with food and eggs until her first offspring appear. After they have mated, several of the daughters return to the nest, expand it, and care only for their own offspring. They defend the nest together, however, and form a family community without any further division of labor. *Halictus ligatus*, a very commonly found North American species, displays similar behavior.

The social development of *Halictus subauratus* goes significantly farther. These shiny, metallic halictids with reddish-yellow hairs on top are roughly 7 to 8 millimeters long and can be found on sandy or clay soils all over the warmer regions of central Europe. The brood cells of these bees branch off the main tunnel but can be surrounded by a system of cavities, as well. The mother outlives her first brood, which grow into exclusively female bees indistinguishable from the mother. They devote themselves to building and collecting. The nest mother continues laying eggs without abandoning the nest. In autumn, she and the "summer females" die, but both females and males appear from the last eggs. Only the mated females survive in their winter hiding place to start the brood process over again the following spring. The relationships among members of *Lasioglossum pauxillum* are very similar. These black-brown, only 5-millimeter-long bees can frequently be found on solid clay soils, on dry lawns, and also in gardens. In the spring, the overwintered female builds an initial nest tunnel not even 3 millimeters wide with up to 25 often grape-shaped, grouped cells. Once these are provisioned with food and an egg, the female closes the nest from the inside and waits for her offspring to hatch. They become clearly smaller female workers, which open the nest and expand it with a system of branching tunnels and new cells. These workers perform all of the jobs in the nest except laying the eggs, which the old nest mother continues to do. The diggings excavated by expanding the nest pile up in layers at the exit, creating an ever-growing chimney as many as 5 centimeters tall. A female guard constantly watches the nest entrance and checks with her antennae to ensure that every returning bee belongs to the nest community. From July on, males and larger females appear. Once the old mother's death seals the fate of the nest, the mated young females start new nests the following spring. As a result of their overlapping generations and division of labor, the previous two examples can be classified as eu-social communities.

Among members of *Lasioglossum zephyrum* that frequent North America, the female—occasionally two or three females together—starts a nest in spring or expands an available previously existing nest. She builds branched tunnel systems without considerable cavities around the brood cells. Each of the founding females apparently digs and provisions only her own cells. The mothers are still in the nest when the young hatch. It is sometimes claimed that *Lasioglossum* mothers feed their larvae in open cells. However, it is only certain that they sometimes open their cells to inspect them. The first offspring are entirely female, smaller, and differently colored than their mothers, which gives the impression that they are a different species. They are also sexually underdeveloped. These daughter-workers carry in food and build and provision cells, while the mother (or mothers) lay(s) eggs. At the end of summer, the eggs hatch some males and some females capable of reproducing. These females mate with partners from their own nest or from other nests. The old mothers and their sons die, while the young females search for a place to overwinter in and build a new social group in the following year.

The rare New-World bee *Evylaeus marginatus* (also called *Lasioglossum marginatum*) is another step up the ladder to an insect colony. In this species, the founding female must remain alone in the nest to feed her offspring in their open cells until they pupate, since she doesn't provide them with provisions ahead of time. Only female bees hatch initially, which are indistinguishable from their mother. They stay in the nest unmated, though, as workers. The community doesn't disband in autumn, but instead lasts four to five years, during which many generations of workers replace each other. In contrast to the nest mother (queen), who lives several years, the daughter-workers live only one solid year. Males are hatched in addition to females only in the community's last summer. The males break into other nests to mate with their females, which form the basis for new nests in the following year.

Halictids are just one of several bee families to show the beginnings of more developed communities. The same social relationships can be found in the large group of carpenter bees, within the subfamily *Xylocopinae*. In addition to *Xylocopini* (large carpenter bees) that gnaw mostly in tree trunks, this subfamily includes the tribe of *Ceratinini* (dwarf carpenter bees) and *Allodapini*, which nest in hollow plant stems. Large and small carpenter bees live alone all over North America, with the exception of their inclination to form overwintering communities. Small carpenter bee females are known to repeatedly open their cells—divided with hardened plant pulp—inspect their brood, and (perhaps) feed the larvae and remove their excrement. Thus, the mother behaves in a sub-social manner. Mainly present in North America, South Asia, and Australia, the *Allodapini*—with their genera *Allodape, Braunsapis, Exoneura*, and others—are also on the way to forming colonies. The nest-founding female lays her eggs in the spacious hollow of a plant stem but not in separate cells. Instead, she lays them next to each other in a community cell. Thus, the hatched larvae lie together in the same room.

They develop within the group, steadily fed by the mother first with soft food from her crop, then with solid pollen. In this way, bodily contact arises between the mother and her young, as with *Evylaeus marginatus*. Once they are fully developed, several of the sterile females remain in the nest to work on the nest and feed the larvae. Others mate and leave to start their own nests or return so that sometimes numerous bees capable of laying eggs are in the nest. Thus, we can observe small, sterile daughters and large, fertile daughters working together with the founding mother. When the mother dies, her offspring—mated and unmated—cooperatively perform all the work of the nest. *Allodapini* are apparently just beginning to learn to divide work among themselves, since they have not developed any significant caste differences among themselves other than size. The nests also remain small and include only a small number of bees (see Figure 5.7).

Orchid bees (*Euglossinae*)—classified in the family *Apidae* near bumblebees (*Bombinae*)—are also advancing toward becoming true social animals. These bees of tropical South America have shiny, metallic green, blue, red, and purple exoskeletons and a tongue that is up to 30 millimeters long, which they carry stretched under their bodies while flying. The genus *Euglossa* is the most well known of these. They build nests that often contain hundreds of vertical, nut-shaped cells thickly covered with resin and placed next to each other in the open or in cavities lined with resin. The resin contains particles of wax (a characteristic product of true *Apidae*) that the females excrete under the last

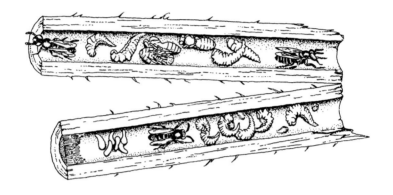

FIGURE 5.7

Nest of the *Allodapini* bee *Braunsapis
sauteriella* in a hollow stem with eggs,
larvae, and larvae as old as first pupae (but
without cocoons). Worker bees provide the
food, which is placed near the larvae in the
form of small pollen clumps.

(*Wilson,* The Insect Societies, *Cambridge,
Massachusetts: Harvard University Press,
1974; after an example from Sara Landry.*)

tergum. *Euglossa* include solitary species as well as species that live together in the same nest. Of these, some young females leave their birthplace to start their own nest. Others remain to build, to lay eggs, and to provide the brood with food. It is surprising that those who inform the others about the location of resin do not help building the nest. Some females provision only "their own" cells, while others provision more generously. The females that depart to start their own nests are somewhat larger than those that stay in the nest, which points to a nascent dimorphism. Nevertheless, *Euglossa* represent a relatively low, quasi-social, stage of development.

Unique Mating Behavior

Orchid bees have another astoundingly peculiar characteristic. The males possess enormous, sack-like, swollen lower tibiae. In contrast to nearly all other male bees, they single-mindedly pollinate flowers, though their purpose is not to gather food. They search for certain kinds of orchids from which they take not nectar but an aromatic essence secreted on the spot of the flowers' atrophied nectaries. The male orchid bees dab this perfume with the help of hair tufts on their front feet, and as they rise into the air their strenuously working legs transfer the essence into the flagons on their hind legs. On the inside of these containers, fine, feathery hairs absorb the aromatic liquid. In this way, a male orchid bee roughly as big as a bumblebee can transport up to 30 cubic millimeters of aromatic essence. The purpose of this is even more peculiar than the method of collecting. This behavior exists exclusively to attract females ready to mate. At first, the scent attracts other males present at the mating site. Each one looks for a blind that he defends against competing males and uses as a base for making repeated courtship flights in which he displays his impressive, colorful armor. Since all the males do this, a real feast of color arises that attracts the females just as much as the scent. When it comes to mating, the mutually circumspect orchid bee males are unabashed competitors.

Carpenter bee males exhibit similar quarrelsome behavior, as do carder bees (*Anthidium manicatum*) males when they defend their highly individual mating territories. In contrast, most digger bee (*Andrena*) males court with more civility. They mark sections of ground with secretions from their mandibular glands. Then, females ready to mate fly into these territories and attract the males' attention with sexual scent secretions. Amazingly, some *Ophrys* flowers of the orchid family also attract the males with a scent that is similar to *Andrena* females' scent. While trying unsuccessfully to mate with these flowers, the misled males pollinate them. Eucerine males of the anthophorid tribe *Eucerini* also allow themselves to be deceived in similar fashion. Mating territories—delineated flying areas where the sexes meet—also exist among bumblebees, stingless bees, and honeybees. It should be noted that the waiting males are generally peaceable among themselves.

The act of mating is not always a simple thing. For this purpose, leaf-cutting bee males possess fringes of branched hairs of different lengths and specialized scent glands on their complexly-built front legs. When the male grabs his partner from behind in flight, his spiked front feet push her head down and flatten her antennae in the grooves of his widened tarsi, simultaneously releasing his sexual scent. With his fringes covering his partner's eyes, his middle legs clutching her wings, and his hind legs bending her abdomen into the air, finally nothing more stands in the way of their coupling.

Due to the diversity of bees, we can expect to find yet more distinctive mating rites about which we know something but certainly far from everything. Every species has its own mating places, and, depending on the species, the rituals take place either in the air or on the ground. Such impatient suitors exist among digger bees that they cannot even wait until the females have emerged from the ground. These males dig the females out themselves. Among all bees, the male mates on top of the female, and the coupling event is quickly over. Mating frequently means the end for males, which lose their lives through the violent loss of their reproductive structures.

Threats from Outside, and Enemies Within the Ranks

It is continually astounding how carefully bees arrange their brood nests, how circumspectly they select their nest sites, how purposefully they choose their building material, how conscientiously they line and disguise their nests, and how they conceive of such architectural finesse. Why do they do all this?

Certainly the lavish inner architecture and wallpapering have something to do with protecting the food stores from moisture. If the pollen-honey provisions become saturated with water, they are easily exploited by fungal hyphae and rot. Silk wallpaper and linings of leaves are good preventative measures. So is the refined strategy of *Dasypoda hirtipes*, who don't simply store their food balls on the ground but instead place them on hardened pedestals of pollen. The ventilated cavity that some halictids build around their brood cells serves the same purpose. But what is the purpose of the piles of soil and tunnel-like external nest structures of some ground- and wall-nesters? Why the barricades in front of nests within plant stems or snail shells? Why the carefully constructed walls around nests built in the open? It doesn't take much guessing to answer these questions: all of these things help to keep out enemies.

Bees are mainly threatened by other insects, such as flies and beetles that parasitize larvae, the predatory checkered beetle (*Trichodes apiarius*), and especially the cunning wasp species like the cuckoo wasps of the families *Chrysidadae* and *Sapygidae*, which break in to the brood chambers of the working bees and attempt to smuggle in their own eggs. If the wasps are successful, their parasitic larvae wait until the bees grow into larvae, then eat them. Ichneumonids are especially dangerous. They do not even need to gain entry to bee nests in order to deposit their eggs on or in bee larvae. Instead, they penetrate the nests from outside with their long egg-laying borers. Another ichneumonid, *Leucopsis gigas*, has even been reported as capable of penetrating the rock-hard wall of closed cells of the mason bee of the walls *Chalicodoma muraria* with its egg-laying apparatus.

Bees are also not safe from enemy attacks when they visit flowers. One threat is the crab spider (family *Thomisidae*), which sits motionless inside a flower until it grabs a landing bee with lightning speed, paralyzes it with a poisonous bite, and finally sucks it dry. The bee wolf (*Philanthus triangulum*), a species of digger wasp, is another threat. The bee wolf female attacks bees busy collecting and anaesthetizes them with one sting. This hunter then carries her prey under her body into a brood chamber at the end of a tunnel about 40 centimeters deep in the ground. Once she has collected three to six paralyzed bees there, she lays her egg on them and closes the tunnel to the chamber. The hatched larva nourishes itself on the muscles of the live, preserved bees.

All the protective measures of bees are useless against the reproductive cycle of the blister beetle (*Meloe*). In spring, this roughly 2-centimeter-long beetle, which secretes a foul-smellling oil when touched, lays up to 2000 eggs in the ground. Very active, first-stage larvae at most 3 millimeters long hatch from these eggs and clamber up the stems of flowers. They then lurk in the blossoms, waiting for bees.

Triungulin larvae, named for their claws divided into three parts, climb onto bees' hairs and hold tight, getting a free ride into the bees' nests. There, they eat the bees' eggs first and then the pollen stores, molting as they do this.

Then, the plump, 1- to 1.5-centimeter-long second-stage larvae crawl into the ground and develop into grub-like third-stage larvae. After pupating in the chrysalis stage, these larvae become adult blister beetles the following spring.

Despite all these threats, the greatest danger to bees comes from their own relatives. Solitary bees, which do not have their own nests or food stores, break into other bees' nests just after the nesting bees have finished making a pollen cake. The solitary bee then lays her egg on the provision. The nest mother appears not to notice the intrusion and lays her own egg in the same cell before she closes it unsuspectingly. Typically, the solitary bee larva hatches first and eats the provisions, so that the nesting bee larva, hungry as it is, must starve to death. Otherwise, the parasite behaves much like its victim would have. At most, it takes somewhat longer to pupate. It can afford the extra time since it has to wait until the next year for more prepared provisions anyway.

Of the 20,000 bee species, roughly 15 percent are cuckoo bees, named after the deceptive tactics of the well-known bird. The idle female cuckoo bees do not possess an apparatus for collecting pollen—they do not need one—and have no thick hair covering, which would only hinder their smuggling attempts in foreign nests and their potential confrontations with nest mothers. The males, on the other hand, do not need to forego an ample covering of hair. Cuckoo bees are found among all bee families, but they mostly victimize related species, as do other parasitic bees like the cuckoo megachilids *Coelioxys*—characterized by their cone-shaped, extended abdomen—and *Stelis*. Both of these are megachilids and parasitize other megachilids (leaf-cutting and mason bees). Some cuckoo bees, such as the parasitic halictid *Sphecodes albilabris*, only victimize one species. This halictid only parasitizes the plasterer bee *Colletes cunincularius*. Other parasitic bees take advantage of a diverse number of species. Those committed to just one or a few species often closely resemble their victims and, amazingly, even frequent the same flowers. However, parasitic bees can also be very different from their victims. They are often brightly colored, which, from the point of view of biological usefulness, is not easy to understand.

The abundance of parasitic bees is evident form the fact that the large bee family *Anthophoridae* includes an entire subfamily of numerous parasitic species. The subfamily *Nomadinae*, often described as "wasp bees" because of their wasp-like appearance, prefer to parasitize a very different family of bees, the digger bees. In contrast, the parasitic *Melecta* only thrive by victimizing members of their own carder bee family.

We can assume that cuckoo bees evolved from non-parasitic species. If they don't reveal their victims, we can gather knowledge about their origins from their developmental rhythm and bodily characteristics. The question remains, however, How did parasitic behavior develop at all? We can imagine that a sudden mutation (through a genetic leap) or gradual atrophy of the collecting apparatus led these bees to their parasitic behavior. We can just as well imagine the reverse: that they lost their collecting apparatus and hair covering after they began a parasitic lifestyle. This question remains unanswerable, like the cause and origin of nearly every other evolutionary occurrence in the animal and plant world.

chapter 6

Bumblebees and Stingless Bees

Despite the enormous number of solitary and budding social bees in the world, they remain unknown to most people, passing their lives too hidden from our view. But everyone knows bumblebees, even though this group is not really that numerous. Worldwide, there are only 400 species of bumblebees, more than 50 of which are present in North America—some of them threatened with extinction. They make themselves powerfully known through their size, their shaggy hair covering, their conspicuous coloring, and their unignorable buzzing (hence their Latin genus name *Bombus*).

Stingless bees, in contrast, live only in the tropics and subtropics, reaching northern Mexico. They include about 200 species, which mainly belong to the two genera *Melipona* and *Trigona*. Their lack of a functional stinging apparatus is their most conspicuous characteristic and also gives them their name.

Bumblebees and stingless bees are not related to each other. Nevertheless, they are introduced here together because they both occupy a position on the social ladder between the wild bees' preliminary stages of social development and the superstar of the colony-building insects, the honeybee. Bumblebees, however, with their relatively small communities and less elaborate nests, occupy a lower place than the stingless bees. A proportion of the latter build large colonies with ornate combs, and their social behavior nearly equals that of honeybees.

Both bumblebees and stingless bees secrete wax for use as building material.

Bumblebees

Despite their threatening, low-pitched buzzing, bumblebees are very peaceful creatures who very seldom use their considerable stinger. They really only become uncomfortable for people to be around when we go in the immediate vicinity of their nests or massively threaten a single bee.

Bumblebees are distributed over the entire world. In Australia and New Zealand, where they were not native, they have naturalized. They appear to thrive best in the temperate zones of the Northern Hemisphere and in mountainous regions. A few species live in tiny colonies north of the Arctic Circle. The thick, furry hairs of bumblebees display the most varied color patterns, with brown, yellow, black, red, and white markings. Unfortunately, the color of the hairs only sometimes corresponds to a species identity, since the coloring is not always species-specific and can vary among males and females and queens and workers within the same species. In addition, color variations exist from locality to locality, as exhibited by by *Bombus crotchii* or *Bombus rufocinctus* in North America, and by the small garden bumblebee (*Bombus hortorum*) in central Europe. We can determine even less about individual bumblebees from their size because dwarves and giants fly out of the same nest at different times. To identify species of bumblebees, you have to examine more constant characteristics such as head shape and the composition of mouth parts (tongue length). Of course, nesting habits and occurrence in certain habitat types can also be helpful. The recent attempt to subdivide *Bombus* into more genera (e. g., *Pyrobombus*, *Megabombus*, *Alpinobombus*) does not make species identification any easier.

Some bumblebees nest underground, where they take over found cavities such as mouse or mole tunnels and the nest-building materials they contain. Bumblebees enlarge these cavities according to their needs—sometimes to lengthen the entrance area or to make the cavity bigger when it becomes too small for their nest. Bumblebees that exclusively or predominately nest underground include: *Bombus terrestris* (large earth bumblebee), *B. lucorum* (small earth bumblebee), *B. subterraneus* (shorthaired bumblebee), *B. soroeensis* (Ilfracombe bumblebee), *B. ruderatus* (large garden bumblebee), *B. mastrucatus*, and more. Yet others, such as *Bombus muscorum* (large carder bee),

B. hypnorum, B. agrorum, B. humilis, and *B. ruderarius,* nest above the ground. These bumblebees select sites on mossy ground, under tufts of grass, and on embankments, as well as in tree cavities, holes in walls, and occasionally in nest boxes, from which a stubborn and persistent queen bumblebee seeking a nest site can sometimes chase off an equally interested pair of titmice. Other bumblebees have apparently not yet decided through the process of evolution whether they should nest in or above the ground. Among these are: *Bombus pratorum* (early-nesting bumblebee), *B. lapidarius* (stone bumblebee), *B. sylvarum* (shrill carder bee), *B. hortorum* (small garden bumblebee), *B. jonellus* (heath bumblebee), *B. monticola,* and *B. pyrenaeus.*

Let's follow one of the most attractive bumblebees, the shrill carder bee (*Bombus sylvarum distinctus* VOGT), through the course of a year. The queen of the shrill carder bee is 16 to 18 millimeters long, making her a moderately large bumblebee. Her thorax is black-brown on top and rimmed with bright brownish-yellow, while her abdomen is striped with alternating horizontal rows of black and light gray and tipped with an orange point. She also has a long proboscis. She occupies previously existing nests underground or builds her own nest above ground. The *sylvarum* female emerges from her winter hiding place under moss or loose grass no earlier than the end of April, when the spring sun has already well warmed the surface of the earth. Unlike some of her cousins, she is in no great hurry. Other bumblebees such as the underground-nesting large earth bumblebee (*Bombus terrestris*) and the above-ground early-nesting bumblebee (*Bombus pratorum*) already appear in the middle of March. After emerging, the shrill carder bee queen satisfies her hunger while searching for nectar in late-blooming willows and already opulent spring flowers. At the same time, she looks for a suitable nest site. Soon, she will find one on the rim of a ditch between a field and a meadow, where a slight overhang provides a certain amount of protection from rain. Despite her Latin name (*sylvarum*, relative to the forest), she stubbornly avoids the forest. She smoothes the ground at the nest site and then collects hay, pieces of moss, dry leaves, and animal hairs (if she can find any) from the vicinity, building a nest bowl similar to a bird's nest. At the entrance, she forms a pot-shaped container 2 centimeters high and 1.5 centimeters wide out of wax: the honey pot. She exudes the wax in the form of small sheets from between the chitin plates on the bottom and top sides of her abdomen. With her hind tarsi, she takes the sheets of wax and places them in her front legs, then in her mandibles, which she uses to knead the wax into a form suitable for construction. The shrill carder bee queen then fills the honey pot with nectar, which serves as food reserve for bad weather and as food for the expected first offspring. Soon after that, she carries fat balls of pollen back to the nest in the "baskets" on her hind legs and mixes them with nectar and her own glandular secretions to make a small ball in the middle of the nest. On this base of nutritious protein and carbohydrates, she erects a wax ring, somewhat smaller than the honey pot, and lays 6 to 12 eggs inside it. Then she closes the brood cell with a wax covering (see the top left of Figure

6.1). When the tiny larvae hatch after three to five days, they begin to eat their own bed, so to speak. Since the provisions are quickly consumed, the queen places pockets stuffed with pollen on the sides of the nest. While the larvae eat into these pads of food, the queen—when she is not flying or working on the nest—sits on the nest and incubates the larvae like a bird. Just as some birds in brooding plumage have a sparsely feathered spot on their abdomen, the queen bee has a thin spot on her abdomen that allows her to transmit warmth more easily. Soon bumps emerge on the queen's cradle-like abdomen from the ever-growing larvae. Approximately eight days after the larvae hatch, they separate from each other, and each one spins a solid, egg-shaped cocoon of silk around itself.

The queen dismantles the wax covering and uses the wax to build new brood containers directly above the old cocoons. Seven to ten days after they pupate—roughly 19 to 22 days after the eggs are laid—the first offspring hatch out of the comb (the collection of cocoons under the wax is referred to as a "comb," even though it pales in comparison to the comb of honeybees). The young female bumblebees are smaller than their mother because of their merely adequate food supply. After hatching, they are still a nondescript grayish-white, but soon take on their beautiful colors. Above all, young *sylvarum* females are infertile. After their hairs have dried and they have eaten heartily from the honey pot, they take over all the jobs of the nest: building cells, bringing in food, warming the brood and providing them with provisions. Egg-laying is the only job they do not

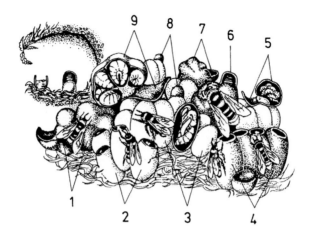

FIGURE 6.1

Nest of a pocket-maker bumblebee.

Top left: After leveling and padding the base, the queen places a honey pot in the nest entrance and the first brood cradle above a clump of pollen in the middle of the nest.

Main image: View of the working nest. 1 Old brood area with a pollen pocket still covered with wax; 2 old empty cocoons serve as honey containers; 3 cocoons with worn out wax covering, one cut open showing a pupa; 4 true honey containers; 5 young larvae in the communal storeroom with pollen pocket; 6 new brood cradle with eggs; 7 new brood area with pollen pocket; 8 new brood structure above exposed cocoons; 9 old brood area with pollen pocket.

(Michener, The Social Behavior of Bees: A Comparative Study, Cambridge, Massachusetts: Harvard University Press, 1974; after an example from J. M. F. de Camargo.)

perform. They are servants to the mother, who now steps out of her initially solitary life and truly earns the name "queen." She also continues to busy herself with nest work, though she does not fly out of the nest anymore. Soon, the second generation of bumblebees hatches. As the population of the colony grows, the queen increases her egg production and puts the growing number of eggs up above the old brood areas. Figure 6.1 provides a view into the growing nest. Compared to the dwarf-like bumblebees hatched first, the following broods are conspicuously larger due to the industriousness of the female workers. *Bombus sylvarum* does not hoard pollen but instead stores honey in several honey pots or in old, empty cocoons.

The members of the colony create more space according to the degree that the colony grows. They do this by pressing the nest material outward and sticking the covering back together with a layer of wax and tree resin. The workers carry the resin—so-called bee glue (propolis)—back to the nest on their hind legs, much as they carry balls of pollen. Lastly, a waxy covering with air holes is built, which makes it easier for the bees to maintain a constant nest temperature of 86 to 90 degrees Fahrenheit—usually significantly higher than the outside temperature. The cold-blooded bumbles (like all bees) create this warmth by moving flight muscles. They can also uncouple their wings and legs. Occasionally, the nest must be aired out, which bumblebees accomplish by fanning their wings.

To defend the nest, the colony assigns guards that remain near the nest's entrance. They fend off small mammals as well as insect enemies and foreign bumblebees. For this job, they need their stinger. Some bumblebees are also able to spray their venom at enemies many centimeters away. The nest jobs appear to be distributed equally among colony members of all ages, except for collecting, which is done primarily by the larger bees. Worker bumblebees die after living six to eight weeks.

Sexually mature bumblebees, first the males, hatch in high summer at the earliest. They emerge from unfertilized eggs, as is customary among Hymenoptera. This form of reproduction is called parthenogenesis. It also occurs in other insect groups—plant lice, for example—and in the lower forms of crabs. It is not totally certain whether, in addition to the queen, one or more workers produce such eggs. The especially well-fed, fertile females that appear last rival the queen's size. While the males initially laze around the nest, fattening themselves on nectar and then abandoning their birthplace forever, the freshly-hatched females can remain many days in the nest. They even make themselves useful by working before they leave to find a mate. In the meantime, the males have been leading the life of a vagabond, spending the night in flower sepals. But they have also done something in preparation for their potentially imminent coupling: they have laid out scent paths—circular courses 150 to 200 meters in diameter that they repeatedly fly in the same direction, making stops at the same plants every time to leave scent marks from mandibular gland secretions. These flight paths vary according to the bumblebee species and can be in treetops, at the height of shrubs (our shrill carder bee), and close to the ground. When a queen of their species circles their area, males wildly throw themselves at her. If, in the exceptional case, males do not quickly appear, the queen waits with buzzing wings until partners arrive. To mate, the male straddles the queen on a solid foundation (not in the air, as with honeybees). The queen normally mates with multiple males one after the other, which die shortly thereafter. Even males that don't manage to mate do not live longer than four weeks. The pregnant female, however, stands only at the beginning of her life's purpose. She finds a suitable overwintering site in a hole in the ground, in moss, or under a rotten tree stump, and spends the cold season in rigid hibernation, living off her fat supply with a much-reduced metabolism.

The old bumblebee colony begins to decay as soon as the sexually mature bees leave the nest around August. No more workers are born, the old ones die outside working or inside the nest, and the queen dies with them. So ends the life cycle of *Bombus sylvarum distinctus* VOGT, the shrill carder bee.

Of course, there are all kinds of species-specific variations among bumblebees. One very apparent variation is in the way the brood is fed. Shrill carder bees, with their long proboscis, provision their brood with pockets of food that they attach to the outside of every brood structure. The growing larvae eat their way right into the provisions. We classify this type of feeder as a "pocket maker." Bumblebee species with a shorter proboscis are generally "pollen storers." These species initially fill cocoons empty of pupae with pollen and sometimes substantially lengthen these storage vessels with wax cuffs. But they don't let the brood feed themselves. Instead, they feed the brood with a regurgitated mixture of pollen and honey through holes the workers have bitten in the brood cells. For this purpose, these species also frequently leave the larval bed open. There are indications that the pocket makers switch to this form of feeding when their queens are developing. In other words, it is possible that something about the change in food quality causes the development of queens. In addition, the possibility that secretions from head glands are also a determining factor should not be excluded. Besides shrill carder bees, pocket makers include the following species: *Bombus pascuorum*, *B. hortorum*, *B. muscorum*, *B. pomorum*, and others. Pollen storers include *Bombus hypnorum*, *B. lapidarius*, *B. terrestris*, *B. pratorum*, and *B. jonellus*. Both types of feeders include roughly equal numbers of species.

The yearly cycle can also vary among different species of bumblebees. For example, *Bombus pratorum* (the early-nesting bumblebee) and *Bombus hypnorum* experience the high point of their development by the end of June, while *Bombus lapidarius* (stone bumblebee) colonies don't fully develop until August. Differences also exist in the ultimate colony size of various species. In high summer, *Bombus sylvarum* colonies include 80 to 150 bees and do not grow any bigger. Other bumblebee species such as *Bombus lapidarius*, *B. hypnorum*, and *B. magnus flavoscutellaris* create colonies of 400 or more individuals. Large earth bumblebees (*Bombus terrestris*) achieve nests with up to 1000 residents. In the far north, bumblebee nests stay very small. Due to the short summer, several species that live on the edge of the polar ice pack—the largest bumblebees and the most thickly covered with hair—seem to want to go back to a solitary existence. They are also not helped by the fact that they look for food in the light of the midnight sun and even in sub-freezing temperatures. The arctic bumblebees are miniature furnaces who manage to raise their muscle temperature in just a few minutes from several degrees above freezing at rest to more than 86 degrees Fahrenheit while flying.

Bumblebees are the most "industrious" insects. In temperate latitudes, they depart the nest at dawn with temperatures under 46 degrees Fahrenheit (queens can fly when it's under 39 degrees Fahrenheit), and they fly until the outbreak of darkness. In comparison, our honeybees "sleep late" and have a much shorter daily routine.

So, what else distinguishes bumblebees from the more highly developed honeybees? Bumblebees do not share information about their sources of food with each other. In other words, they do not possess a "language." There is also no mutual exchange of food among bumblebee adults. Their colonies last only one year, and this does not change in southern climates without cold winters, where the pregnant females start new nests without hibernating, or return to their old nests to temporarily share authority with the old queen.

People have often tried to distinguish plants favored by bumblebees. There are various grounds for this. Though these thickly furred bees mostly have a broad spectrum of flowers at their disposal, their very long probosces lead them to plants with especially deep-lying nectar. Bumblebees are the preferred pollinators of the pea, labiate, and figwort families. Red clover, beans, vetch, and other cultivated legumes are absolutely dependent on them. Some flowers, like snapdragons, appear to actually invite their visits. Bumblebees do not avoid any such complex flowers. Monkshood, larkspur, dead nettles, and sage are visited almost exclusively by bees of the genus *Bombus*. They penetrate the deep blossoms of foxglove and bellflower, which honeybees seldom accomplish. But there are also specialists among bumblebees that find an especially comfortable way to the deep-lying nectar of some flowers. They simply bite holes on the outside of the flowers' flanks, "robbing" the nectar and defeating the biological point of their visit.

Like solitary bees, bumblebees also include parasitic species. They belong to the genus *Psithyrus* and can be called cuckoo bumblebees. *Psithyrus* females penetrate the nests of *Bombus* bumblebees, lay their eggs inside, and let the workers of the host nest raise their offspring. Appropriately, this parasitic species does not include workers, just males and fertile females. These do not possess a pollen basket on their hind legs, and they cannot produce wax.

Every *Psithyrus* species parasitizes either one specific *Bombus* species or just a few. *Psithyrus vestalis* prefers the nests of large earth bumblebees, whom they strongly resemble. In contrast, *Psithyrus sylvestris* parasitizes a species they do not really resemble, the early-nesting bumblebee. The vast majority of parasitic bumblebees do look like their hosts, though, even if their wings are usually not so beautifully amber-colored but instead somewhat smoky looking. They also manage to move their wings more quietly. This doesn't accomplish much, however, since bumblebees probably do not detect sound waves, anyway.

When possible, parasitic bumblebees sneak unrecognized into young nests with only a few workers or, if the nest is busier, they attempt to enter with a quick, directed approach. Sometimes "bloody" conflicts with the nest residents occur, which the *Psithyrus* bumblebees, with their thicker armor and more powerful stinger, usually win. As soon as they take on the scent of the nest, they are tolerated. In the best situations, the invader and the queen of the nest get along side by side. The *Psithyrus* female often oppresses the queen, however. When she doesn't sting the queen right away, she eats the queen's eggs or prevents her from laying more. The development of the *Bombus* bumblebee colony ends, and the parasitic offspring hatch and abandon the nest to mate. The *Psithyrus* males then die, and the females overwinter. When they leave their winter quarters the following spring—somewhat later than their hosts—and select a host nest, the "tragedy" begins all over again.

Among bumblebee species there are also some individuals that occasionally "forget their nature" to become more comfortable. These queens could have been equipped with everything they need to start a nest. But for some reason or another—perhaps they awoke from their winter hibernation too late or found no well-suited nest site—they break into an already started nest of the same or a different species, kill the reigning queen, and take over the nest. This leads to mixed nests that can produce two different bumblebee species. People often attribute such behavior to *Bombus lucorum* (small earth bumblebees), who seek out large earth bumblebees (*B. terrestris*) as hosts. Science calls this behavior facultative (incidental) parasitism.

Stingless Bees

If we climb one step higher in the world of eu-social bees, we come to a group of bees classified by their lack of a stinging apparatus. The ancestors of these bees certainly possessed a sting that has now withered away. Stingless bees are not defenseless, though. Especially species that live in large colonies can be very aggressive. If a curious person comes too close to their nest, he can be badly injured. The disturbed bees hurl themselves at him and bite him in the skin and hair, holding on tightly. In addition, some species spray a sticky, acrid secretion from their mandibular glands into the bite wounds that can leave permanent scars. Thus, fire bees or hair-cutting bees are fitting names for these species.

As *Meliponinae*, stingless bees are part of a subdivison of the large family *Apidae*, just like bumblebees. With over 300 species, stingless bees are the largest subfamily. Although they have recently been divided into a number of differently named genera, it makes sense for our purposes to focus on two of the tribes in the old system of classification: *Meliponini* and *Trigonini* (meliponines and trigonines). All stingless bees live in the tropics (with only a very few in the subtropics). Meliponines live exclusively in the New World. The extensive trigonines are native to both the New World and Africa, as are members of the predatory genus *Lestri-melitta*, which we will simply classify as *Trigonini*.

While meliponines often strongly resemble our honey-bees in size and appearance, trigonines differ widely among themselves in both size and the thickness of their hair covering. Especially tiny trigonines are not even 3 millimeters long. Whether their frequently used name "sweet bee" is appropriate becomes doubtful, at the latest, if they attack you, covering your bare skin in massive numbers and pene-trating your nose, mouth, and corners of your eyes.

Stingless bees include both primitive species that form small communities with few members and species (espe-cially among trigonines) that form large colonies with up to 80,000 individuals. Sometimes stingless bees build their nests in the open, but they mostly build in protective cavi-ties such as hollow trees, holes under rock piles, holes in the ground, and—in emergencies—abandoned termite nests. Their building material is wax, which they secrete on the upper side of their abdomen. They seldom build with it alone, however. Instead, they mix it with resin (propolis) that they carry to the nest in baskets on their hind legs. This brown mixture is called cerumen. Most stingless bees surround their nest with a many-layered cover that encloses a narrow, hollow cavity: the involucrum. This cavity serves to stabilize the nest temperature and protect nests built in the open from enemies. A thick, insulating upholstery made from a mixture of wax, propolis, and clay or chewed-up plant material fulfills the same purpose. It is often placed on the rim of the nest or—in cavity nests—above and under, or in front of and behind the nest, respectively. The hard material in this protective barrier is called batumen.

Several primitive stingless bee species simply put their brood cells together in a disorderly pile. The majority of species, however, build well-ordered combs of hexagonal cells. These wax combs consist only of horizontal layers of cells arranged parallel to each other, like the paper combs of yellow jackets. The bees build them from bottom to top, letting small cerumen pillars maintain the necessary gaps between the layers. Some species spiral their combs into the air. Depending on the size of the nest, the middle comb layers can achieve the circumference of a soup plate. Only one exceptional stingless bee, the African *Dactylurina staudingeri*, builds combs like honeybees, double-sided and vertically constructed from top to bottom. Unlike bottom-opening wasp cells, the cells in stingless bee combs open on top. Thus, the growing larvae and pupae do not hang with their heads downward like wasp larvae and pupae, but instead stand upright in the cells. Figure 6.2 shows an example of a higher meliponine's nest structure.

Stingless bees collect provisions. They store honey and pollen in their own containers on the margins of their nests. Pollen is kept in high, slim containers; honey is stored in bulbous pots often as big as pigeon eggs. The way stingless bees rear their young strongly resembles the way solitary bees rear their young: First, the brood cell is provisioned

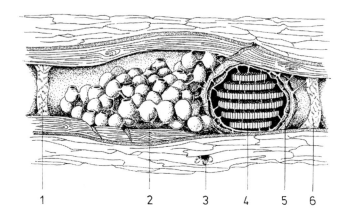

FIGURE 6.2

Nest of the stingless bee *Melipona interrupta*.
1 Batumen barrier, 2 containers for provisions,
3 entrance, 4 brood combs, 5 involucrum,
6 batumen barrier.

(*Michener,* The Social Behavior of Bees:
A Comparative Study, *Cambridge, Massachusetts:
Harvard University Press, 1974; after an example
from J. M. F. de Camargo.*)

with a mixture of pollen and honey; then, the queen lays an egg on the provisions, and the workers close the cell. However, some stingless bees do not content themselves with a one-time provisioning of their offspring. They feed them directly as soon as the larvae have consumed the first provisions. Only after this do they close the cells. Before the young bees hatch out of their cocoons, the workers gnaw off the wax covering and other recoverable wax remains to use in other ways. Thus, the cells are only used as incubators once, and the combs must be steadily renovated.

Workers and males develop in equal-sized cells, with the males hatching out of unfertilized eggs parthenogenetically. Among trigonines, females destined to become queens are allotted somewhat larger cells. Their development into queens can simply be explained by the fact that they receive more food because their cells have more space. Workers and queens develop in the same-sized cells among meliponines. Since they appear to receive equal amounts of food, scientists find it difficult to explain how some meliponines develop into queens. We suspect that genetic factors are responsible for the creation of the two different castes. The oversupply of potential young queens must be reduced later. Apparently, a special food is occasionally fed only to the genetically determined queens, which leads them to look and behave like workers. This happens when the colony's environmental conditions and food income are not favorable and additional workers are needed. When the colony is well nourished, the bees simply kill the surplus queen genotype individuals.

Stingless bees surprise us with yet another peculiarity. The workers of almost all social bees lay eggs (unfertilized, of course) under certain, mostly difficult, conditions for their colony. Stingless bee workers, however, lay eggs all the time. Their egg cells do not have a nucleus. These so-called trophic eggs are used to feed the queen. Meliponine workers lay trophic eggs on the prepared larval food in the brood cells, where the queen finds them, eats them, and then lays her egg. Trigonine workers deposit especially large, yolk-rich eggs on the edges of the brood cells before they fill the cells with food. The workers in both genera of stingless bees can also lay male eggs capable of developing into bees. We assume, in fact, that the majority of males in meliponine and trigonine nests descend not from the queen, but from workers.

While we simply recognize the size difference between worker and queen bumblebees to distinguish them, stingless bees have other, clearer distinguishing characteristics. The queens do not possess pollen baskets on their tibiae and thus cannot collect pollen. Internally, they are missing wax glands. The queens' head, eyes, and wings are also substantially smaller than those of the workers, but the queens have a larger abdomen. Though meliponine queens are not larger than workers when they hatch, trigonine queens are bigger from the beginning.

To describe stingless bees, we can justifiably speak of two differently developed castes and use the customary term for this: dimorphism. A division of labor also exists between the two castes, in which the queen lays eggs and the workers take care of all the colony's remaining needs. Normally, only one queen is responsible for producing female offspring, but often numerous potential queens can be seen within the same colony. Since sexually mature females are born throughout the brood period, more and more fertile princesses stay in the nest. After hatching, however, they must hide in the involucrum or in the outskirts of the nest, to avoid being killed by the workers. The survivors either encounter a possibility to start a new nest, or they wait to take over the reigning queen's job when she dies.

A division of labor has also been observed among workers. Younger workers busy themselves in the nest and soon begin building (because of their early developing wax glands). Their older sisters concentrate on collecting building materials and food. Nest work such as building, brood rearing, and cleaning is divided among the workers that remain in the nest independent of their age.

The "stay-at-home bees" probably also share guard duties. One guard is enough in small nests with a narrow entrance. There, the guard creates a front door with her head, so to speak. Some species close the entrance over night. Others build a cerumen tube in front of the entrance or use long entry tunnels in the ground. Still others build cuffs around their entrances out of sticky propolis, especially to keep out ants. Some large colonies fashion wide vestibules in their entryways many centimeters long, wherein a well-armed guard stands ready. Particularly aggressive meliponines who nest in the open secrete an alarm substance from their mandibular glands when threatened. They use this secretion to put each other in a heightened attack mode.

At least a few colony-rich species of stingless bees have developed a communication system that allows pollen foragers to alert their sisters at home to abundant sources of food. The successful returnees move excitedly on a zigzag course through the nest. This stimulates some of the colony members to fly out of the nest. They know what to look for based on the floral fragrance the successful collector brings to the nest with her. The high-pitched buzzing produced by the wings of foragers among some trigonine species during their hasty runs of encouragement through the nest probably also signal something to their nest mates, who sense the vibrations and are stimulated by them. Other species have actually invented a method to direct the way to a food source. The foragers produce an aromatic secretion in their mandibular glands, which they leave on stones, stems, piles of soil, or similar objects in intervals of 2 to 3 meters on their way back to the nest from a food source. The alerted young foragers in the nest really only need to follow this fragrant path to reach the food source. Astonishingly, the successful foragers also guide their nest mates to the food. First, they fly back and forth near the nest to draw the attention of the other waiting foragers, who then stick hard on the heels of their leader.

Since they have lost their solitary instincts, meliponine and trigonine queens are no longer able to start a nest on their own. They require the support of workers. As soon as the sexually mature female offspring appears in the nest, a small troop of worker "pioneers" starts looking for a new nest site. If they are successful, they quickly begin building the nest and food pots. They bring building materials in their pollen baskets and food (in the form of a honey-pollen mixture) in their crop from their home nest. Once the foundation for the new nest is laid, a small swarm with a young queen follow the "pioneer" workers into the nest. More workers (and occasionally other queens) follow. They begin to build brood cells, and they ensure that only one queen remains alive. This queen then leaves the nest to look for a mate. Numerous males ready to copulate wait for her at gathering spots in the air near the nest. Mating occurs in flight and probably only once. Only after mating does the young queen finally settle into the new nest. After a few days, she begins laying eggs in the provisioned cells.

Stingless bee colonies last many years. Because they exist in southern regions, they do not require winter pauses in activity. Theoretically, one colony could survive for an unlimited amount of time, since a young queen can immediately replace a dead nest mother.

Many kinds of small mammals and occasionally people threaten stingless bees. When there were no honeybees in the Americas, the Mayans welcomed stingless bees as economically useful. In Brazil, native Indians sometimes still keep trigonines in small wood and clay containers. The harvested honey keeps poorly and is sour and thin. Since honeybees are now distributed over the whole world and are of incomparable economic use, stingless bees play essentially no economic role today.

Stingless bees have no parasites among their relatives. They are, however, not without related nuisances. One species, *Lestrimelitta limao*, has developed into a robber in South America and Africa, although the African species are grouped under the genus *Cleptotrigona*. Instead of tediously flying from plant to plant collecting pollen and nectar, this species gains provisions by violently entering peaceful bee colonies. They have adapted themselves so completely to robbing that their workers do not possess any pollen-collecting devices anymore. They carry their plunder in the form of a honey-pollen mixture back to their own nest in their crop. They even tear off wax material from other nests and take it with them. This species appears to have originally developed out of trigonines, but they do not just rob this genus. They also attack meliponine nests and occasionally even honeybee hives.

To attack, the robbers disperse a lemon-like scent inside a foreign nest, which makes it difficult for the attacked bees to distinguish friend from foe. Still, these attacks can become embittered battles using mandibles and "chemical weapons," or they gather into clashes that persist over weeks, in which case the defending colony generally falls by the wayside. It not only loses its provisions but is also often destroyed.

chapter 7

On the Summit of Social Insect Life

Bees of the genus *Apis* have reached the highest form of social life. Although honey is also the indispensable food of all other bees, only representatives of this genus are called honeybees. They collect this treasured food in such quantities that they do not just satisfy their own needs with it but also generally leave some extra for greedy humanity.

Honeybees fulfill all the criteria for a eu-social society: a colony includes only overlapping generations of one family, the fertile and infertile females differ in appearance (dimorphism), a division of labor exists within the worker caste, and they have the ability to communicate messages about worthwhile food sources using motion signals (which stingless bees also intimate). In addition to this so-called physical communication, honeybees also possess chemical means of communicating—their glandular secretions, which are important for the communal life of a colony.

Currently, scientists distinguish nine species of honey-bees. All of them build two-sided, vertically oriented combs with regular hexagonal cells in which they store provisions and raise their brood. Bee researchers generally accept that honeybees originated in south or southwest Asia. Even today, all honeybee species live exclusively in Asia except for our western honeybee.

The Genus *Apis*: Species and Races

Some pronounced differences exist among the nine honeybee species. Body size is one of these. Unlike in bumblebees, size is a relatively species-specific characteristic in honeybees.

Figure 7.1 shows the comparative sizes of seven *Apis* species. In order to gain an overview of the individual species' lifeways, it is prudent to divide them into open nesters and cavity nesters.

Four species are open nesters. By this, we mean that they build only one comb, which they constantly fortify to protect their provisions and brood. They include two giant species and two dwarf species. The first giant species is called *Apis dorsata* and has long been simply dubbed giant honeybee. It is found in tropical Asia, including Indonesia and the Philippines, chiefly in humid coastal regions. These hornet-sized bees with rusty-brown hairs and smoky-colored wings build an impressive, up to 1-meter-tall and 1.5-meter-wide comb that is usually widely visible. They fortify their comb with the heavy branches of tall trees or construct

A size comparison of seven of the currently known honeybee species (approximately 3/4 life size).

Top left: The giant honeybee of the Himalayas, or Nepal honeybee (*Apis laboriosa*).

Bottom left: The giant honeybee (*Apis dorsata*).

Top center: The western honeybee (*Apis mellifera*).

Center: The eastern honeybee (*Apis cerana*).

Bottom center: The cavity-dwelling bee (*Apis koschevnikovi*).

Top right: The little honeybee (*Apis florea*).

Bottom right: The small dwarf honeybee (*Apis andreniformis*).

it under cliff ledges (see the left drawing in Figure 7.2). As a relatively primeval form, they utilize the same cell type for storing provisions and for rearing workers and drones, though the drone cells are covered somewhat higher. Like in all other *Apis* species, giant honeybee queens emerge from cone-shaped structures on the bottom edge of the comb.

Often, numerous colonies nest together. Large colonies include as many as 40,000 bees. With the onset of the dry season, they wander over long distances in moist mountain forests. Their restlessness thwarts every attempt to domesticate them. Giant honeybees become aggressive when disturbed and temporarily abandon their comb. Indian "honey hunters" find it worthwhile, however, to get into the nests with a lot of smoke at night. One comb contains up to 40 kilograms of honey.

The giant honeybee of the Himalayas (*Apis laboriosa*) is even somewhat bigger than the giant honeybee and lives in cool, high valleys of the Himalayas and in the mountains of west India between 1300 and 4100 meters elevation. Giant honeybees of the Himalayas have a thick, light-brown hair covering. Their nests under cliff ledges are usually difficult to access, and the huge comb can grow up to 1.5 meters tall. In the cold season, the colonies select overwintering sites at lower elevations between 1200 and 2000 meters, where they gather together without a comb in a grape-shaped ball. During the roughly six-week overwintering period, the temperature inside this ball of bees can sink to just a few degrees above freezing.

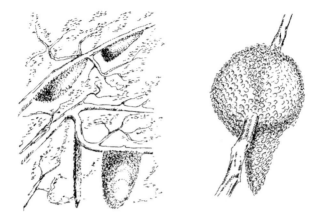

FIGURE 7.2

Left: Single comb nest of the giant honeybee (*Apis dorsata*).

Right: Comb of the little honeybee (*Apis florea*). Its upper part has rings widely extended around a weak branch.

Little honeybees (*Apis florea*) are among the smaller open nesters. They are distributed all over the lowlands of southern Asia, like the giant honeybee, and extend even farther west to the Persian Gulf. In contrast to *Apis dorsata*, they can survive in the driest and hottest regions. A little honeybee is no bigger than a housefly. With its gold-orange abdominal rings and thick, bright, felted hair covering, it is a feast for the eyes. The comb of little honeybees is no bigger than a soup plate, and these bees like to hide it in partly shaded bushes. As a result of its extended honey cells, the upper part of the comb widens nearly into a ball that encircles a supporting branch (see the right drawing in Figure 7.2).

The bees affix a ring of tree resin on both sides that they keep soft and sticky to protect the comb against ants. As with all other honeybee species (except for the giant forms), little honeybee workers and drones develop in cells of similar shape but different sizes. The number of bees in a nest is limited to between 6000 and 11,000. *Apis florea* drones exhibit a unique characteristic on their hind tarsi: a thumb-like projection that may play a role in helping them clutch a queen during mating. When disturbed, a colony abandons its nest site and finds another one instead of defending itself. These bees are not very settled, anyway. Their colonies easily swarm and duplicate themselves. This frequent resettling gave them the name "nomad bees." The small nests of little honeybees do not provide people with much honey. Their combs, however, have commercial value in Thailand. They are sold and eaten together with their brood.

The small dwarf honeybee (*Apis andreniformis*) was once typically considered a subspecies of *Apis florea*, whose lifestyle it mostly shares. The small dwarf honeybee is perhaps a bit smaller and not so distinctive as *Apis florea*, though, with darker coloring and a somewhat less uniform appearance. Small dwarf honeybees behave more aggressively when disturbed. They are distributed throughout the countries and islands surrounding the South China Sea.

First among the cavity-nesting honeybee species is the eastern honeybee (*Apis cerana*), also often called the Indian honeybee, which is distributed over all of central and eastern Asia, including the Indian and Japanese islands. These bees are housed in hives and used for economic purposes. Their colonies are smaller than those of our western honeybee, with 10,000 to 20,000 members. They also only build five to eight side-by-side hanging combs. Members of *Apis cerana* possess the unique ability to move their entire colonies as "swarms" to new, more favorable locations when they are repeatedly disturbed or lacking food. They closely resemble the western honeybee but are somewhat smaller, at least in southern areas, and exhibit great color variations between black and yellow depending on their geographically determined subspecies. In contrast to our honeybees, they also possess four hair bands on their abdomen instead of three, and they stand with their abdomen toward the nest entrance (instead of with their head toward it) when a draft of cool air comes into the hive. Eastern honeybees do not use resin as building material, and they leave a small hole open in the middle of the raised cappings on their mature male brood. Compared to the western honeybee, they are also more peaceable and seldom use their sting. They try to scare off enemies by rapidly moving their wings, which creates a characteristic sibilant sound. One of their most dangerous enemies is the powerful hornet *Vespa mandarina*. When this hornet ventures near the entrance of an *Apis cerana* colony, the bees wait patiently until the hornet sits to deal with it instead of flying immediately toward it, which would mean their death. They do not sting the wasp but instead suffocate it with the heat generated inside their nest (as hot as 113 degrees Fahrenheit).

The cavity-dwelling bee *Apis koschevnikovi* strongly resembles *Apis cerana* in size, body characteristics, and behavior. Its whole appearance seems reddish, even though its metanotum and the underside of its abdomen are more brightly golden-yellow. Like *Apis dorsata*, its wings are somewhat smoky-colored. *Apis koschevnikovi* was only recently rediscovered as a unique bee species and appears to be limited to the tropical rainforest of Borneo and Sumatra, where it nests in hollow trees. Colonies of these bees can also be kept in hives. They are good-natured, but like *Apis cerana* bees, they have a strong penchant for swarming and easily abandon hives. If their drones did not have a different daily time for flying than *Apis cerana* drones, their status as an independent species would be doubtful. The males of the two species also have different sexual organs. *Apis koschevnikovi* bees (like *Apis cerana*) have a flight and collecting area of only just over 500 meters, smaller than that of our western honeybee and *Apis dorsata*, which disperses over a 2- to 3-kilometer radius.

Let's now focus on the western honeybee (*Apis mellifera*), which we've already referred to numerous times as "our honeybee." Strictly speaking, its Latin name is not quite correct, and the name's origin is regrettable. In older German bee literature, the name *mellifica* appears everywhere. However, the term *mellifera*, which was always common in Anglo-American literature, is now globally used. Thus, the "honey-making" bee (*mellifica*, from *facere*, to make) became the "honey-carrying" bee (*mellifera*, from *ferre*, to carry), which is wrong because bees do not collect and carry honey—they collect and carry nectar. They make honey only once they have thickened the raw material and mixed it with glandular secretions. But since an internationally prescribed principle, the priority rule, states that the initial name should always take precedence when an organism has many current names, scientists thought they must settle on *mellifera*, without considering that Carolus Linnaeus recognized his own mistake and rectified it with *mellifica*.

The western honeybee is the only one of the nine honeybee species that oriented itself toward the west in geologic history. We assume that *Apis mellifera* separated itself from the eastern honeybee at the beginning of the Ice Age approximately 2 to 1.5 million years ago. After it settled in Europe and Africa, it presumably was cut off from its place of origin by the growth of deserts. In Europe, bees went as far west as they could and penetrated far into the north. In Africa, they made it all the way to the southern tip. *Apis mellifera* accompanied European conquerors and colonists into all areas of the New World.

The wide natural expansion of honeybees makes it no surprise that subspecies which display all kinds of differences in appearance and behavior have developed in various regions. The fact that different authors described such races in their native places once led to many mix-ups and much uncertainty. New developments in the measurement of external form (morphometry), such as the use of numerous individual characteristics, and modern biometric and molecular genetic technologies have only recently made exact classification and subdivision possible. Based on these developments, we now believe that the western honeybee should be divided into at least 25 bee races. Most of them are native to tropical and subtropical regions. Fourteen of these races live on the Mediterranean coast alone.

Of course, we cannot describe the location, characteristics, and behavior of all the *Apis mellifera* subspecies here, but we will at least present three of the most important races for beekeepers worldwide:

1. *Apis mellifera mellifera*, the dark, or northern, bee, a vigorous bee with a covering of long brown hairs and two attributes valued by apiarists, namely a relatively late onset of brood in spring and a moderate temperament;

2. *Apis mellifera carnica*, the carnica bee, a slim, peaceable bee with a gray-hair covering, long proboscis, and the ability to develop large, efficient colonies astonishingly quickly out of relatively small overwintering groups;

3. *Apis mellifera ligustica,* the Italian bee, an especially beautiful bee with a shining, yellow abdomen that overwinters in large colonies, broods early and heavily, and distinguishes itself through its particularly gentle disposition.

We assume that today's honeybee races, including the three just named, developed only relatively recently in Earth's history, during the Pleistocene Ice Age 15,000 to 10,000 years ago. During this period only a narrow, inhospitable belt of tundra existed in central Europe between the northern and southern glaciers, from which most of the plants and animals, including the bees, had to retreat toward the Mediterranean Sea. In the various coastal regions, different races of bees developed in accordance with the local living conditions. As the glaciers retreated to the north about 10,000 years ago, and flora and fauna moved northward again, the dark bee migrated from the French Mediterranean coast north through the Alps into all of central and northern Europe, and as far east as the Ural Mountains. The carnica bee, which may have developed on the Dalmatian coast, moved over the northern Balkan Peninsula into the Danube valley, up to the Carpathian Mountains, and westward into the valleys of the eastern Alps. In contrast, the Italian bee never left its place of origin, since it apparently never succeeded in crossing the Alps.

That is the natural distribution of these three bee races. Only recently have bee breeders caused considerable movement and much range overlapping among them. Today, all three of these bee races can be found in North America.

The Domestic Honeybee

Assuming our contention that honeybees have the most highly developed colonies is correct, we can intellectually construct the stages of their development and find hints in the societies of living wild bees. Admittedly, however, contemporary bee species cannot offer us any more knowledge about the ancestors of honeybees than that. The social wild bees that exist today are not immediate ancestors of honeybees. Such ancestors certainly existed, but they now must be extinct. Today's semi-social and social bees are simply branches on the phylogenetic tree of bees.

The organization of domestic honeybee colonies is so fascinating that many naturalists consider it justified to call them "superorganisms." A bee colony consists of a number of single insects, just as an animal or plant organism consists of many single cells. All colony members constantly relate to each other in great harmony. They react as a unit when they must differentiate their colony from other colonies or respond to the most varied demands of their environment. All of a bee colony's achievements equal those of a single, more highly developed animal.

The Comb

Our honeybees raise their brood in cavities and prefer hollow trees or other protective nest sites when left to their own devices in the wild. In such a cavity, honeybees build a parallel arrangement of regularly shaped combs. Their building material is a soft, malleable wax that they secrete as small, transparent plates from glands under the wax mirrors on the underside of their abdomen. They remove the wax plates from their bodies with the tarsi of their hind legs and transfer them, with their middle and front legs, to their mandibles. There, they knead the wax and mix it with glandular secretions to prepare it for use. The wax producers are female. They hang together tightly like a curtain under the growing comb and regularly pass their building material up to the wax-handling workers on the comb. What emerges there is a wonder of precision, functionality, and aesthetics. The comb consists of a vertical middle wall, from which slightly tilted, upward pointing, hexagonal cells extend on both sides. Three cells join below the base of each cell, dividing the floor into three rhombus-shaped areas (see Figure 7.3). With the exception of the thick outer comb edges, the cell walls are only a tenth of a millimeter thick. The average cell diameter from wall to wall is 5.37 millimeters in worker cells and 6.91 millimeters in drone cells. Only the cradles used for brooding queens seem out of place. They are hanging bowls that, once closed, resemble acorns or peanuts. Since these structures are artless and only used once, it seems possible that they are atavistic remnants from a long-ago stage in the development of honeybees. The bees rear their brood in the combs and nearly as often in the cells. The cells also serve as storehouses for provisions. Pollen is always stored close to the brood, with honey following it. Beekeepers call this "organization of the comb."

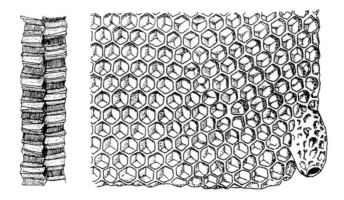

FIGURE 7.3

The marvel of the honeybees' comb.

Left: Lengthwise.

Right: Overhead view with queen cell on the side.

People have often wondered why bees build their combs the way they do and not otherwise. We have come to the conclusion that the sole reason is economical and effective use of space. A 30-square-centimeter section of delicate comb can hold 1 kilogram of honey.

Appropriate natural honeybee nest sites have become so rare that in Europe, almost no bee colonies are found in the wild anymore. In the southern United States, however, africanized honeybees nests are quite common. People make artificial nest sites (hives) made of straw, wood, or plastic available for bees. The bees take advantage of this help and also please the beekeeper by building their combs in wooden frames that make it easier to extract the honey.

In addition to wax, honeybees occasionally process tree resin, known as propolis (from the Greek *pro*, before, and *polis*, city) among beekeepers. Like bumblebees and sting-less bees, they shave this sticky substance off the buds and trunks of resin-rich trees with their mandibles and carry it back to their hive in the baskets on their hind legs. They fill in small gaps and cracks in the hive walls with this propolis and use it, mixed with wax, to close larger holes, such as the hive entrance for the winter. Honeybees also use propolis to embalm large invaders, like shrews or the honey-robbing death's head hawkmoth, that do not manage to escape from the hive.

The Colony and Its Individuals

At its summer high point, one honeybee colony numbers approximately 60,000 to 80,000 members. They are almost all female workers, with only a couple of hundred drones (males). Only one individual, the queen, is a fertile female and simultaneously the mother of all colony members.

The three honeybee types exhibit noticeable differences in appearance (see Figure 7.4). The queen is one and a half times bigger and twice as heavy as a worker. She possesses an especially long abdomen, which contains prolific, paired ovaries, both containing 180 egg tubes. (Bumblebee queens only have four egg tubes, meliponine females only 6 to 10!) An important part of the sexual organ is the seminal vesicle, in which the sperm supply for the queen's entire lifetime is stored. She receives this supply during a few mating flights shortly after hatching. The queen does not have a pollen-collecting apparatus, wax glands, or scent glands like the workers. Instead, she issues commands that are indispensable for the social life of the colony using other glands, unique to queens. The drones are distinguished by their abdomen, which is plump, blunt-ended, and covered with thick hairs at the back. They are equally big as the queens. Their large, round head with powerful mosaic eyes is also conspicuous. Like all males of the order Hymenoptera, they do not possess a sting, a pollen-collecting basket, wax glands, nor anything that could be remotely construed as useful for work. As is customary among Hymenoptera males, drones hatch from unfertilized eggs. This means that

FIGURE 7.4

The three types of *Apis mellifera*.

From left to right: Queen, drone, worker.

the queen can stop the entry of sperm when she lays drone eggs. Thus, the parthenogentically produced drone eggs only carry maternal genes and, in contrast to the queen and workers, possess only half a set of chromosomes.

The "lazy" drones—which are fed by the workers as long as they stay in the hive—live only a couple of summer weeks, regardless of whether or not they fulfill their sole purpose: to mate with a future queen. These "summer bees" live only as long as the drones. You could say that these bees would work themselves to death for the benefit of the colony if a natural law did not predetermine the length of their lifetime. The "winter bees," born in autumn, do not have to care for the brood; therefore, they can accumulate a thick padding of fat in their abdomen from which they live for many months. This is important because in winter there are no offspring. The ability to overwinter as a colony distinguishes the genus of honeybees from all other bees. A queen lives up to five years. She thus survives numerous generations of her offspring, guaranteeing the continued existence of the colony. In the summer, the queen undergoes an intensive metabolism, which enables her to lay up to 2000 pin-shaped eggs weighing 0.13 milligrams and measuring 1.5 millimeters long. This is then her only active job. The workers devotedly clean her and feed her while comprising a constantly changing "court" around her (see Figure 7.5). They feed the queen an especially protein-rich food, which they produce in substantial glands (lateral pharyngeal glands) in their heads.

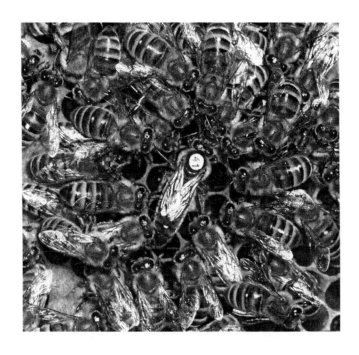

FIGURE 7.5

The queen is surrounded by a court of
workers that continually clean and feed her.

Brood Rearing and Division of Labor

The business of rearing brood in a honeybee colony begins in early spring and lasts until late autumn. Each larva, less than 3 millimeters long, hatches from an egg laid by the queen on the bottom of a cell after three days. Nurse bees immediately surround the larva with a sort of milk secreted from their head glands. This protein-, fat-, and sugar-rich food supplies nutrients to the offspring during their first three days alive. After that, they receive a coarser mixture of honey and pollen. Only larvae that are meant to become queens continue to receive the liquid food until they reach the pre-pupa stage. Their food, called "royal jelly," is not exactly the same as the liquid food given to worker and drone larvae, though it is not fundamentally different, either. The almost identical ingredients are simply combined in different amounts in the two forms of liquid food. There is no "miracle food" that people long believed accounted for the development of queens.

After six days, a worker larva weighs 500 times more than when it hatched. Once the larva becomes elongated, the worker bees close its cell with a wax cover. Twelve days after it was laid as an egg, the elongated maggot spins itself into a fine cocoon and pupates. Queen maggots do this two days earlier, drone maggots two days later. While worker bees require 21 days for their complete development, queens need only 16, and drones develop in 24 days.

As soon as a young worker bee leaves her cell, she is bound into the multi-faceted work of the colony. She hardly has time to clean herself or arrange her still-moist covering of hair before she begins a three-week-long stint of internal service, which begins with cleaning the empty brood cells for a new generation. After that, she feeds older larvae pollen and honey. Once she has consumed a lot of pollen herself and developed glands to produce liquid food, she provisions young larvae with the nutritious brood milk. Meanwhile, she also becomes a part of the queen's court. About eight days after she emerges from the cocoon, the young worker bee takes nectar from returning collecting bees and passes it on to other hive bees. While it is transported, the nectar is thickened and enriched with enzymes, then stored in cells in the form of honey. The collecting bees deposit pollen they have gathered in the cells themselves. Then the hive bees pack it in tightly with their heads. The glands that produce liquid food gradually disappear on young worker bees, and wax glands begin to develop. After they have lived 12 to 18 days, worker bees become builders and subsequently defenders of the colony. They patrol the vicinity of the hive entrance, looking for any moving object that does not belong to the hive and attacking it when necessary. At the age of 21 days, worker bees become collectors. From then on, their sole job is to import honey and pollen into the hive and carry in, if necessary, propolis and water.

Such is the course of a worker bee's life, as dictated by age. Extraordinary situations can, however, disrupt this work plan. Occasionally, younger bees must fly out of the hive to collect pollen, and older collectors must reactivate their wax and liquid-food-producing glands to take on the jobs of younger bees. In an emergency, this ability to change jobs can be crucial for a colony's survival.

What Holds a Colony Together

Food exchange does not only occur between collecting bees and hive bees. It also occurs constantly among workers that stay in the hive. Scientists call this "trophallaxis." This exchange does not serve to quiet hunger, since every bee could easily take what it needs from the colony's provisions. Instead, the bees pass on information about crucial events in the life of the hive during trophallaxis. For instance, when hive bees only reluctantly take nectar from returning collecting bees but voraciously accept deliveries of water it sends the signal: bring more water quickly! During hot days, the hive temperature is lowered with water.

A honeybee colony is nothing without its queen. In her absence, the workers quickly become restless, their desire to work lags, and social cohesion unravels. Such "queenless" bees promptly attempt to brood a new queen. If young worker larvae are still present, the workers widen their cells into queen cradles and try to induce them to become queens by feeding them royal jelly. Beekeepers call this occurrence "emergency queen rearing." If it does not work, the worker bees start laying eggs themselves—drone eggs, naturally— which unfortunately does nothing to alleviate the colony's disastrous situation. The downfall of such colonies is usually certain.

The queen is the regulatory mechanism of a colony's social fabric. How does she do this? People long wondered what a queen's especially large upper mandibular glands could be good for, until we discovered in the 1930s—primarily through the research of the English scientist C. G. Butler—that these glands produce a substance crucial to the cohesion of a colony. The primary effect of this "queen substance" comes from the major components 9-oxo-2-decenoic acid and 9-hydroxy-2-decenoic acid. Glands under the chitin on the queen's abdomen are also somewhat responsible for her harmonizing influence on the workers. The queen spreads these highly effective substances over her body with her legs. Then the court bees eagerly lick them off and pass them on to all hive residents with their mouth partas and antennae.

Honeybees are only satisfied when enough queen substance is present in their hive. If this substance is missing, emergency queen rearing occurs. When the amount of it decreases as the queen grows older, the worker bees rear a substitute queen while the old one is still present. They build a queen cradle on the middle of a comb in which the dying queen lays one of her last eggs. This is called supersedure. If, on the other hand, the supply of queen substance only becomes low because the colony has grown substantially over the course of a summer, the bees prepare to "swarm."

The queen substance is not only effective within the hive, but also outside of it. For example, its aromatic components ensure that a bee swarm in the air stays together instead of scattering in different directions. Young, unmated queens also issue a message with their scent, steering potential mates toward them.

It seems curious that a queen's glandular secretions, rather than her bodily presence, regulate all of the important events in a bee colony. These secretions—or more precisely, the combination of secretions—belong to the biological substances we call pheromones. They work much like hormones, but outside of the body instead of inside and only among individuals of the same species. The fragrance (a mixture primarily comprised of geraniol, citral, and farnesol) produced by workers in a gland on the tip of their abdomen is also a pheromone. They spread this fragrance by stretching their abdomen upward, uncovering their Nasonov glands, and vigorously whirring their wings. This attracts their sisters when they want to show them where a rich food source is, or when they want to make it easier for inexperienced young workers to find the hive entrance. Pheromones such as alarm and deterrent pheromones excreted from mandibular glands and glands near the stinger also play a role in defending the colony against large enemies. These pheromones prepare the workers to attack.

How New Colonies Originate

A bumblebee queen can and must start her nest alone. A honeybee queen is not suited for this. She does not overwinter alone but with her entire colony, which is substantially smaller in the winter, numbering only about 10,000 individuals. To overwinter, the bees come together into a crowded grape-shaped ball. The temperature on the surface of this ball does not go much below 50 degrees Fahrenheit. In spring, when new provisions are brought in, the bees begin to brood at a constant temperature of 95 degrees Fahrenheit, and the size of the colony grows rapidly. This growth is not unlimited, but before the hive becomes too small, the colony divides itself. For this purpose, the bees position small, downward-opening queen cradles on the edges of the combs, in each of which the queen lays one egg. As soon as the young larvae begin to elongate and the first of these queen cells is covered, the queen leaves the hive with a portion of the colony in a prime swarm. After a short period of buzzing about, this swarm settles on a branch or at another suitable place, retaining its grape-like shape. It soon moves again, however, to a new nest site that scouts have already searched out. There, the swarm first builds combs, then collects food, broods, and promptly begins a new existence. In the colony they left behind, the young queens hatch in the meantime—one after the other, if possible, to avoid encounters, since they fight each other mercilessly if the occasion arises. Each queen then leaves the hive with a following of bees, which are called afterswarms. Usually, a small number of bees remain in the original hive, and the last young queen takes over.

While the old hive queen in the prime swarm can quickly continue laying eggs, the young queens, which, we will assume, find suitable nest sites with their accompanying bees, still have to mate. So, they make a quick succession of mating flights. Generally, mating occurs in the air at selected sites, so-called "drone congregation areas," in the surrounding area. Each drone only copulates once, then dies. After mating multiple times, the queen returns to her colony and soon begins to lay eggs. Shortly after swarming occurs, the remaining drones are pushed out of the hive and starve. The war-like story about a slaughter of drones found in many bee books should not be taken literally.

Sensory Capabilities

Honeybees spend most of their time in their dark hive, where they have to rely primarily on their senses of touch and smell. Both of these senses are located on their antennae. The fine sensory hairs, or sensilla trichodea, distributed over the bee's whole body are especially thick on the tip of the antennae. The second segment of the antenna, the pedicel, also contains the Johnston organ, which registers location and body movement. (Since it also measures wind resistance and flight speed, this organ is important for flying, as well.) Bees can detect subtle vibrations in the subsoil. Tones that we hear, such as sounds made by queens before swarming, they feel as tremors on the honeycomb. People long assumed that bees are completely deaf, until it was discovered that they can sense sounds carried by the air with their antennae, even at a distance of only a few millimeters.

The several different types of peg-like structures and round, flat pore plates on the antennae are responsible for honeybees' sense of smell. Every beehive has its own smell, so bees can easily discern between friends and enemies. Bees register chemical alarm signals with their sense of smell, and they use it to perform the important job of identifying flower fragrances. So, it is not surprising that they respond much more strongly to the smell of flowers than they do to any of the technical scents that humans sense more clearly. Honeybees can also "smell" water and sense carbon dioxide, which is not unimportant for their life in a hive, as well as heat and moisture. In order to survive the winter actively, and to keep the brood nest at a constant temperature of 95 degrees Fahrenheit in the summer, heating (through muscle movement) and cooling (through fanning) sometimes become necessary. By bringing in water and spraying it on their combs, honeybees can cool their hive more. Instead of depositing the water on the comb all at once, they can cool even more efficiently by emitting water in tiny drops from their mouth and spreading into a thin film by repeatedly unfurling their proboscis.

The antennae are only partially responsible for honeybees' sense of taste. Taste receptors also exist in their mouthparts and even on their front feet. Bitter things do not bother them. They do not take notice of sugar water, which tastes sweet to us with even a 2-percent sugar content. They only pay attention to sugar water when it contains at least 4 percent sugar, and that only when there is nothing else available, for if honeybees consumed liquids with a low sugar content, they would have to do an unnecessarily large amount of evaporating and thickening to achieve honey's high sugar content of roughly 80 percent.

Fine pads of hair are the receptors for honeybees' sense of gravity, without which no plant or animal can survive. These receptors are located on joint-like connectors between the head and the thorax and between the thorax and the abdomen. Not every life form has a magnetic sense, but honeybees and many birds do. They use it to register the Earth's magnetic field. Millions of tiny, parallel crystals containing iron on the front part of the abdomen are responsible for honeybees' perception of magnetic field lines.

Likewise, not every life form possesses a sense of time. Honeybees have an excellent sense of time at their disposal. They follow the day's 24-hour rhythm without having to rely on the position of the sun; their sense of time also functions if you keep them in an unlit room or transport them over continents. Once trained to come to a food dish at a certain time of day, they will come at that time even in a place in a new time zone. Many plants do not produce nectar all day long but only at certain times, such as in the morning or in the afternoon. Honeybees notice this and thereby save themselves unnecessary collecting flights. Their sense of time is also important because it helps orient them according to the sun's position. We will discuss this shortly.

For their collecting activities, honeybees rely on the vision provided by their large compound eyes. Since these eyes are a grid of individual eyes, we can assume they produce very different images than those we receive through our "camera" eyes. A photograph taken through a honeybee's eyes reveals a mosaic-like view. So, it is no wonder that they can barely discern the details of figures. Instead, they possess a highly developed ability to resolve flickering light. They can detect 250 light stripes per second; we can only detect 40. When sitting still, honeybees can only react to moving objects. Experiments have shown that they cannot discern the difference between simple figures like triangles, squares, and circles when flying, while they can easily distinguish between very sharply divided structures. This kind of vision doubtlessly serves them well when zooming through blooming fields and searching for flowering shrubs and trees.

Honeybees also see colors differently than we do. Their range of vision is shifted toward the short wave end of the spectrum. Thus, they cannot see red, but they can see ultraviolet, which we cannot see. Otherwise, they can distinguish all the colors of the rainbow, though not quite as sharply as we can. Some bright flowers reflect ultraviolet and thus must look very different to bees than they do to us. For example, a poppy that is red for us is not black for bees as we would assume because of their blindness to red. Instead, a poppy appears ultraviolet to bees, however that might look. Some flowers that appear to be one color to us have a spot in the middle that reflects ultraviolet light. Bees are led to the ovaries by this "nectar guide." Another advantage that bee eyes have over ours is that they can detect the polarization of light. Polarized light vibrates in only one plane. This means honeybees can discern, for instance, the variations in polarization of light in different places in the blue sky.

Artists of Orientation

As people began to research how honeybees find their way around in the open, they made startling discoveries. First, we will ask how they take note of certain places such as their hive or a productive food source. A honeybee that leaves her hive for the first time immediately turns around in flight and makes small loops in the air, with her head pointed toward the hive entrance. In this way, she memorizes her hive's appearance before she flies on into the landscape in a straight line. When she returns, she repeats these pendular motions while releasing a scent from her scent glands to mark the hive for inexperienced young workers. (This was once sensible when honeybee colonies were not crowded together under the care of beekeepers, and the unspecific orientation scent, or Nasonov pheromone, could not also draw the attention of bees foreign to a hive.) Through various experiments in which beehives were moved it was discovered that returning bees do not note the hive entrance as such. Instead, they only use it as a destination and rely on conspicuous characteristics of the surrounding area. Likewise, they use identifying features in the area to steer toward crowded gathering places outdoors. To locate a feeding place, such as a flower, a honeybee notes color, structure, and scent. She does not smell the flower until she is in its immediate vicinity, but the scent is easily remembered and makes a stronger impression than the color. This explains honeybees' famous steadfast visitations of the same kind of flower. A collecting honeybee visits one identically scented flower after another, regardless of these flowers' color. It is only important that these flowers belong to the same species—only this guarantees successful pollination.

On long-distance flights over land, honeybees can note landmarks (e. g., streets, river courses, forest edges), but these are only memorized incidentally, so to speak. Their primary orientational reference point is the sun. When a honeybee travels to a distant food source, she memorizes her flight direction's angle to the sun. This helps her immediately find the food source by herself again if, due to a storm or outbreak of darkness, she cannot leave the hive for a long period. Though the sun has moved during this period, the honeybee takes into account the distance she covered. She can do this by using her inner clock. This ability is not instinctive, however; it must be learned. A honeybee only requires a few days to develop this ability, and it's sufficient for her to simply watch the sun's position for an hour or two at a time. She then knows the sun's entire path, even on the other side of the Earth.

Of course, honeybees must also be able to navigate without the sun, and their "sun compass" actually functions when only a small bit of blue is visible somewhere in the sky. The blue sky appears patterned to them because of their ability to sense polarized light. Since the pattern of the sky changes regularly with the sun's movement, honeybees can identify where the sun is and navigate by it simply using this pattern. Even when the sky is overcast, the sun's ultraviolet rays penetrate a thin covering of clouds. Honeybees can see ultraviolet, so they can also identify the position of the sun when the sky is cloudy. Only a thick covering of clouds prevents them from orienting themselves by the sun. Then landmarks must be sufficient, though in this case, honeybees usually do not leave their hive.

The Language of Honeybees

Naturally, honeybees' famous language does not consist of words. It is a form of communication through dance movements. The renowned bee researcher Karl von Frisch deciphered this language. When collecting bees "dance" on the hive's comb, they communicate to their sisters the direction, distance, and extent of the food source they have found outside. Simultaneously, they stimulate other collecting bees to fly to the food source. Figure 7.6 shows the two important dance figures.

If a food source is close to the hive, let's say within approximately 50 meters, then a collecting bee performs a "round dance." She moves in a circle about twice the size of a quarter, then turns around and follows the circle backward. Several bees of collecting age follow her and recognize through the kind of dance and the scent she carries that there is something with this scent to collect nearby. If the food source lies 100 meters or farther from the hive, the collecting bee chooses a different form of dance that signals the source's direction and distance. In this so-called "waggle dance," she outlines a broad figure eight. On the line that connects the two loops, she buzzes her wings and shakes her abdomen swiftly back and forth. This important midsection of the dance communicates the direction of the food source

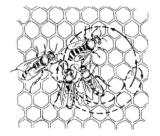 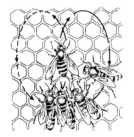

FIGURE 7.6

The honeybee dance.

Left: The bee shows a food source close to the hive with a round dance.

Right: The "waggle dance" signals a distant nectar and pollen source.

from the hive. If the dancing bee points directly upward, it tells her followers to look in the direction of the sun. If she moves with her head pointed directly down, she signals her followers to look opposite the sun. The angle she waggles her abdomen reveals the food source's angle to the sun. If she waggles 60 degrees to the right of vertical, the food source lies 60 degrees to the right of the sun. If she dances 45 degrees to the left of vertical, the food source lies 45 degrees to the left of the sun. Thus, the honeybee possesses the amazing ability to translate a visual experience— namely, the angle between her flight path and the sun—into an experience of gravity: the angle she dances in relation to the vertically oriented comb. Since the dancing bee vigorously moves her wings while waggling her abdomen, we can view the waggle dance as a sort of symbolic walk. During this walk, her wings produce a buzzing tone that her followers can "hear" only over a distance almost tantamount to touching the dancer. Silent dancers do not attract any followers.

The speed of the dance tells the followers the food source's distance from the hive. The faster the dance, the closer the food is. The slower the dance, the farther away the food. The rate at which the dance figures are repeated provides information about the quality of the food source (its abundance, sugar content, and such).

Of course, it would be interesting to know how a honeybee measures the distance she flies to a food source. There are substantial indications that she does this by registering how much food she consumes during her flight, which reveals her energy expenditure. Though occasionally doubts have arisen about this, they do not change the fact that somehow she knows how far she has traveled.

Only bees that closely follow and make physical contact with the dancer on the comb are actually recruited to go to the food source. When they leave the hive, they find the intended food source without any further help or guidance.

This dance language contains many other amazing facets. A honeybee must sometimes overcome a strong crosswind during its foraging flights. Her body's longitudinal axis then no longer lies directly on the path she is trying to fly. Instead, it tends slightly in the direction of the wind. Naturally, the bee then sees that she is at the wrong angle to the sun. In the hive, however, the bee reports the angle to the sun that she actually flew. With her compound eyes, she is capable of determining her angle to the sun within 3 degrees. She is also apparently able to calculate and correct derivations from her intended flight path caused by a crosswind. And what happens, when a honeybee must fly around a high hill to reach a food source? Then she directs the other bees in the hive to fly the most direct route, as the crow flies, that she herself has never flown. A mechanism to account for miscalculation is also necessary in this case, even though it is not easy to understand this work of art's biological purpose. People long puzzled over a small "mistake" found over and over again in the dancer's directions until they discovered that this "misdirection" relates to periodic daily variations in the Earth's magnetism. Thus, this only appears like a mistake to humans, not to bees.

Honeybees do not only communicate about food sources. They also "advise" each other about which shelter they should move into when they are traveling with a swarm. Shortly after a swarm has gathered at a provisional place, some bees begin dancing at the tip of the swarm. These are scout bees that have already been looking for a nest site. Soon more scouts, or "pilots," participate in this dance. Those who have found a good site communicate this in their dance. The language is the same as the language for relating food sources, but the dance takes place on the horizontal top of the swarm and is aligned directly with the preferred destination. At first, the bees do not always agree. Some dance for a distant site to the north, while others dance for a closer site to the south. With time, the bees that make their intention more energetically known (in a quicker succession) win out. We do not know how honeybees judge a nest site's quality. A swarm only breaks up to move into its new nest site when all its members agree on the site. They move into the new place, where many workers start intense scenting to attract their colony mates.

Amazingly enough, there are also different "dialects" in the honeybee language. Different races of honeybees display differences in dance rhythm—especially in communicating distance (dance speed)—and also change from the round dance to the waggle dance based on different distances. The Italian bee, for example, changes from the round dance to the waggle dance when the food source is 35 or more meters away. The dark, or northern, bee does the waggle dance for sources 65 meters away and up, while the carnica bee uses waggle dance to describe any food source 85 meters or more from its hive. Among Italian bees and some other races, an intermediate figure between the round dance and the waggle dance also exists, the "sickle dance." In this dance, the position of the sickle's open side contains a directional component.

Apis mellifera, our honeybee, is not the only bee that possesses a dance language. Other honeybee species dance and use similar figures. Nevertheless, they would not understand each other when put together. The dance rhythm, which provides information about the distance of the food source, is especially different among various honeybee species. Eastern honeybees give each other much more precise information about the location of nearby food sources than our honeybees do, performing a waggle dance for sources only 2 meters away. The large, open-nesting species require a view of the sun or a patch of blue sky to dance correctly on their vertical comb. Little honeybees dance on the wide upper platform of their comb. They did not learn to communicate the angle of the sun with deviations from a vertical axis and do not need to.

A honeybee colony's ability to overwinter and multiply through swarms is enough to show that these insects have achieved a high degree of social development. With their forms of chemical and especially physical communication, they have doubtlessly reached the high point of bees' evolutionary history. No other insect has perfected social life like the honeybee.

chapter 8

Nest Aids
for Wild Bees

Our wild bees are among the world's threatened creatures. The landscape has been covered with houses, industrial estates, recreational facilities, paths, and roads at the cost of wild bees' native habitat. Large, agricultural monocultures also contribute to this destruction of habitat. The use of insecticides not only destroys pests but also kills useful insects, including first and foremost bumblebees and wild bees. In addition, herbicides eliminate plants valuable as food sources for wild bees, sometimes even plants certain bee species depend on completely to survive.

As nature and bee lovers, we are responsible for doing all we can to preserve bees' natural habitat. Those of us who have our own piece of land or a garden are fortunate. There we can cultivate wild plants that, as food sources for wild bees, are becoming hard to find. We can also create sandy clay areas in sunny spots or erect steep clay walls, which many ground-nesters call home. But even if you only have a small backyard or just a balcony you do not need to go without the pleasure of living with wild bees. There are bees seeking nest sites not only in the country but also in the middle of big cities. Of course, you cannot lure the entire native variety of wild bees to your balcony. Some bee species have requirements for climate, location, building material, vegetation, etc. that are much too specialized to live on a balcony. Bees that find their way to us primarily nest in old walls and narrow tunnels (especially hollow plant stems). These include mason bees (*Osmia* species), miner bees (*Andrenidae*), leaf-cutting bees (*Megachilidae*), resin bees (*Heriades* species), and masked bees (*Hylaeinae*). All you have to do is provide a suitable nest site for these bees. Collect some reeds or hollow stems of elder or blackberry brambles, for example, cut small pieces out of them, bundle them, and hang them as a sort of mass accommodation. You could also put them in an empty box or other suitable container to keep them together. Thin bamboo canes, available in garden centers as plant supports, also work well as nest tunnels.

Almost every household has a drill. It is not terribly difficult to obtain a thick wood block (preferably hardwood), a cross section of a tree, an unworked piece of square timber, or an old fence post. To make a nest aid for solitary bees, drill holes 4 to 8 millimeters in diameter 10 centimeters apart from each other in one of these wooden objects. Use the whole length of the drill bit to make the holes as deep as possible, since the bees you want to attract build linear

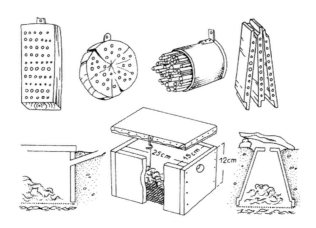

FIGURE 8.1

Several examples of nest aids for
solitary bees (top row) and for
bumblebees (bottom row).

(After Mühlen/Schlagheck, Wildbienen:
Biologie, Bedrohung, Schutz, *Münster,
Germany: Institut für Pflanzenschutz,
Saatgutuntersuchung und
Bienenkunde, 1991.)*

nests. The top row of Figure 8.1 shows several suggestions for such nest aids.

Naturally, you will want to know what is happening in the brood nests and how the young bees are developing inside. To find out, you can extend a portion of the holes and insert glass tubes sealed with wax at one end. Some bees looking for nest sites are not offended by these tubes and build their nests inside them. You can then pull them out from time to time to observe what's happening inside. Such glass tubes are sold by laboratory or medical supply companies.

For wild bees that nest in walls or in the ground, you can fill wooden boxes with sandy clay or loess and position it horizontally, or even better, vertically. You can then poke holes in the soil with a dowel rod or simply hope that the bees will dig their own holes. Dry, sunny places protected from the wind are best for these nest boxes. If necessary, provide a roof to protect the nests from rain.

Of course, other interesting creatures besides bees will appear in such nest aids. Members of the wasp families, such as mason and potter wasps (subfamily *Eumeninae*) and other digger wasps (family *Sphecidae*), should be welcome. All of these bring animal food into their tunnels. Visits of ichneumon wasps (*Ichneumonidae*), cuckoo wasps (*Chrysididae*), and parasitic flies should not be viewed favorably, however. You can also do without ants, which move through the nest area as robbers, and especially without spiders, which themselves become victims of digger wasps.

We can also help bumblebees in their search for a nest site. When people don't clean out bird nest boxes in the fall or early spring, different species of bumblebees often freely move into them. The remains of a bird nest suit these bees perfectly, and by occupying such nests they save themselves work.

For bumblebees that nest in the ground, it is good to bury small wooden boxes in protected places. Each box should have a hinged lid covered with tarpaper. In addition, it should have a short, slanted entrance tunnel 20 to 25 millimeters in diameter made of rolled-up tarpaper. In order to prevent moisture from collecting in the brood area, the bottom of each box should be made with a fine wire mesh, and the box itself should be set on a loose layer of dry leaves or moss. The inside of the box should be lined with a soft material such as wool, down, or moss.

In place of a wooden box, you can also try burying a flowerpot upside down. The nest entrance (the hole in the bottom of the pot) should be protected from rain with a small stone slab. The bottom row of Figure 8.1 shows two examples of bumblebee nest boxes. Incidentally, artificial bumblebee boxes and nest aids for solitary bees can be purchased from zoological supply companies. Ask for addresses in pet stores.

It is important for solitary bees as well as bumblebees that you install the nest aids from March onward. The nest site should always be the same, since bees prefer to return to the location of their birth. By keeping your nest aids in the same place, you also experience the joy of watching solitary bee colonies grow bigger every year. Bumblebee nests do not last through the winter, and you must use the cold part of the year to clean the nest boxes and line them with new padding material. Once a nest box has been utilized, it is very likely that it will be used again. If the young bumblebee queens do not have to fly very far away to find a suitable overwintering site, it encourages reuse. Bumblebee queens are known to spend the winter underground beneath leaves, moss, or rock piles, and in layers of straw or woodpiles. If you have such things in your yard, you should suppress your desire to tidy them or take them away. The bumblebees will be thankful for this.

Abridged Bibliography and Further Reading

Alcock, J. *Animal Behavior: An Evolutionary Approach.* Sunderland, Massachusetts: Sinauer Associates, Inc., 7th ed. 2001.

Bonney, R. E. *Beekeeping: A Practical Guide.* North Adams, Massachusetts: Storey Books, 1993.

Bonney, R. E. *Hive Management: A Seasonal Guide for Beekeepers.* North Adams, Massachusetts: Storey Books, 1991.

Buttel-Reepen, H. v. *Die stammesgeschichtliche Entstehung des Bienenstaates.* Leizpig, Germany: Georg Thieme, 1903.

Carpenter, F. M. "Hexapoda." *Treatise on Invertebrate Paleontology*, Part R, "Arthropoda," Vol. 4. Boulder, Colorado: Geological Society of America, Inc., 1992.

Crane, E. *Bees and Beekeeping: Science Practice and World Resources.* Ithaca, New York: Comstock Publishing Associates, 1990.

Eidmann, H., and F. Kühlhorn. *Lehrbuch der Entomologie.* Hamburg/Berlin, Germany: Parey Verlag, 2nd ed. 1970.

Free, J. B., and C. G. Butler. *Bumblebees.* London: Collins, 1959.

Friese, H. *Die europäischen Bienen (Apidae).* Berlin/Leipzig, Germany: Verlag Walter de Gruyter, 1923/

Frisch, K. v. *Aus dem Leben der Bienen.* Heidelberg, Germany: Springer-Verlag, 10th ed. 1993.

Frisch, K. v. *Bees: Their Vision, Chemical Senses, and Language.* Ithaca, New York: Cornell University Press, rev. ed. 1972.

Frisch, K. v. *The Dance Language and Orientation of Bees.* Cambridge, Massachusetts: Harvard University Press, reprint ed. 1993.

Goetsch, W. *Vergleichende Biologie der Insekten-Staaten.* Leipzig, Germany: Akademische Verlagsgesellschaft Becker & Erler, 1940.

Gould, J. L. and C. Grant Gould. *The Honey Bee.* New York: W. H. Freeman & Co., reprint ed. 1995.

Hubbell, S. and S. Potthoff. *A Book of Bees: And How to Keep Them.* Boston, Massachusetts: Mariner Books, reprint ed. 1998.

Kearns, C. A., and J. D. Thomson. *The Natural History of Bumblebees: A Sourcebook for Investigations.* Boulder, Colorado: University Press of Colorado, 2001.

Khalifman, I. *Bees.* Stockton, California: University Press of the Pacific, 2001.

Michener, C. D. *The Social Behavior of the Bees: A Comparative Study.* Cambridge, Massachusetts: Harvard University Press, 1974.

Moritz, R., and Southwick, E. *Bees as Superorganisms: An Evolutionary Reality.* New York: Springer-Verlag, 1992.

O'Toole, C. and A. Raw. *Bees of the World.* New York: Facts on File, Inc., 1992.

Remane, A. *Sozialleben der Tiere.* Stuttgart, Germany: Verlag Gustav Fischer, 3rd ed. 1976.

Ruttner, F. *Naturgeschichte der Honigbienen.* Munich, Germany: Ehrenwirt Verlag, 1995.

Ruttner, F. *Biogeography and Taxonomy of Honeybees.* New York: Springer-Verlag, 1987.

Sammataro, D., et al. *The Beekeeper's Handbook.* Ithaca, New York: Cornell University Press, 3rd ed. 1998.

Schmidt, G. H. *Sozialpolymorphismus bei Insekten.* Stuttgart, Germany: Wissenschaftliche Verlagsgesellschaft, 1974.

Seeley, T. D. *The Wisdom of the Hive: The Social Physiology of Honey Bee Colonies.* Cambridge, Massachusetts: Belknap Press, 1996.

Sladen, F. W. L. *The Bumblebee, Its Life History and How to Domesticate It.* London: Macmillan & Co., 1912.

Vivian, J., et al. *Keeping Bees.* Charlotte, Vermont: Williamson Publishing, 1986.

Wheeler, W. M. *The Social Insects: Their Origin and Evolution.* London: Paul, Trench, Trubner & Co., 1928.

Wilson, E. O. *The Insect Societies.* Cambridge, Massachusetts: Harvard University Press, 1974.

Index

Figures are indicated by page references in italics.

A

Africa, 99, 106, 115
algae, *13*
Alps, the, 117
amphibians, 3, 10, *13*, 29
angiosperms, 5
animal migrations, 27–28
ants
  army, 3
  colony-building, 24, 31, 33
  Cretaceous origins, *13*
  diet of, 5
  fossils of, 9
  group within Hymenoptera, 3
  and polymorphism, 32
  as robbers of bee nests, 148
apes, anthropoid, *13*
aphids, 16, 22, 26
Apoidea (bee family), 3, 24
Apterygota (wingless insects), 10
Arctic Circle, 83
Arthropoda, 2, 29
  as bee phylum, 3
Asia, 8, 72, 108
associations, family, 28–30
Australia, 9, 72, 83
Australoplatypus incompertus (beetle), 31

B

baboons, 28
bacteria, *13*
Balkan Peninsula, 117
Baltic amber deposits, 6
Baltic seacoast
  source of primeval bees, 6

batumen, 99
bee wolf (wasp), 79
bees
  Allodape (genus), 72
      Allodapini (carpenter), 72–74
  anatomy, 36–*37*, 38–*39*, 40, 42, 46,
           48–50, 54–55, 59, 73–74,
           80, 85, 98, 102
  Andrena (digger), 76
  Andrena jacobi (miner), 50, 65
  Andrena vaga (European willow
      miner), 49, *51*, 76
  andrenid, 38, 40–41, 49–50
    bryony-mining, 41
    comfrey-mining, 41
    female behavior, 49–50
    nests of, 49–50
    species in U.S., 49
  Andrenidae (andrenid), 38, 40, 49,
      146
  Anthidium (carder), 38, 41, 52
    manicatum, 76
    punctatum, 52–*53*, 54
    strigatum, 54
  Anthophora parietina, 60–*61*
  Anthophoridae (digger, carpenter,
      cuckoo), 59–61, 64 81
  Anthophorinae (miner), 38, 40
  Apidae (family), 73, 97
  Apis (genus), 107–8
    cerana, 114
    dorsata, 108, 114–15
    koschevnikovi, 114
    laboriosa, 110
    mellifera, 8, 115–16
      carnica, 116–17

bees: *Apis* (genus): *mellifera* (cont'd)
        ligustica, 117
        mellifera, 116
    See also under bees: honeybees
Ashmeadiella (helicophile), 58
attacks by, 63–64
Augochlora (genus), 66
Augochlorella (genus), 66
Augochloropsis (genus), 66
Augochloropsis sparsilis (halictids), 67
Bombinae (bumblebee), 44, 73
Bombus (genus), 81–95
    agrorum, 86
    Alpino, 85
    crotchii, 85
    hortorum, 03, 86
    humilus, 86
    hypnorum, 85–86, 93
    jonellus, 86, 93
    lapidarius, 86, 93
    lucorum, 85, 97
    mastrucatus, 85
    mega, 85
    monticola, 86
    muscorum, 85, 93
    pascuorum, 93
    pomorum, 93
    pratorum, 86–87, 93
    pyrenaeus, 86
    pyro, 85
    ruderarius, 86
    ruderatus, 85
    rufocinctus, 85
    soroeensis, 85
    subterraneus, 85
    sylvarum, 86
    sylvarum distinctus VOGT
        annual cycle of, 86–88, 90–93
    terrestris, 85–86, 93, 97
    See also under bees: bumblebees
Braunsapis (genus), 72
    sauteriella (carpenter), *74*
bumblebees, 7, 32, 38, 44, 60, 62,
        73, 76, 121, *147*

bees: bumblebees (cont'd)
    anatomical variations, 84–86
    annual cycle of, 86–88, 90–93
    colonies, 90, 92–94
    nesting, 85–*89*, 90, 130, 149
    overwintering, 150
    plants visited, 95
    pocket makers, 92–93
    role of queen, 86–*89*, 90–94, 96
    scent paths, 91–92
    worldwide habitats, 83
carder, 38, 41, 76, 81, 85–87, 92
    female behavior, 52–*53*, 54
    nests of, 52–*53*, 54–55
    species in U.S., 54
carpenter, 36, 41–42, 59–60, 62, 64,
        74, 76
    female nesting behavior, 72–73
Ceratininae (carpenter), 36, 60, 62
Ceratinini (dwarf carpenter), 72
Chalicodoma (mason), 38
Chalicodoma muraria (mason of the
        walls), *57*–58, 59, 79
Chalicodoma sicula (mason of the
        walls), 59
classification, difficulties in, 44
Cleptotrigona, 106
collective action, 63–65
Colletes cunincularius (plasterer), 81
Colletidae (masked; plasterer), 40–41,
        45–*47*, 48–49
colony building, 10, 24–25, 44
    collective tendencies toward,
        63–75
    percentage involved in, 35
    social classifications in, 34
    See also under bees: bumblebees;
        honeybees
combs of, 24, 33, 88, 99, 101,
        108–*111*, 110–*11*, 112,
        118–*20*, 121, 143
    See also under bees: honeybees
communications among, 107,
        138–*39*, 140–43

bees (cont'd)
  competition among, 42
  cooperative care of offspring, 27
  Cretaceous, origins of, *13*
  cuckoo, 59–60, 80–81, 95–97
  cycle, annual, 86–88, 90–93
  Dactylurina staudingeri, 99
  Dasypodinae (melittid), *61*–62
  dependence on flowering plants, 5–6,
          41–42
  Dialictus (genus), 66
  digger, 59–60, 64, 77, 81
  dimorphism, 102, 107
  Eucera longicornis (anthophorid), 65
  Eucerini, 76
  Euglossa (genus), 73, 75
  Euglossinae (orchid), 44, 73–74
  Evolution, relative to genus *Homo*, 10
  Evylaeus (genus), 66, 71
      marginatus (halictids), 71, 73
  Exoneura (carpenter genus), 72
  exotic, 44
  families, *43*–44
  glue. *See* resin
  Halictidae (sweat), 40, 64, 66
  halictids, 40, 65
      development of community,
              69–72
      genera distinctions among, 66
      nesting, 66–*68*, 69–71
      nests of, 66–*68*
  Halictinae, 66
  Halictus (genus), 66
      ligatus, 69
      ongulus, 65–66
      quadricinctus, 69
      sexcinctus, 69
      subauratus, 69
  helicophile, 58
  herbivorous diet of, 5, 35
      adaptation for, 36–*37*, 38–*39*
      sources, 40–42
  Heriades (resin bee), 38, 146
  Heriades truncorum (resin), 55

bees (cont'd)
  holometabolic development, *23*
  honeybees, 2, 7–8, 18, 32–33, 36–*37*,
          38–*39*, 40, 42, 49, 76, 76
          92, 94–95, 99
      colonies of, 110, 118, 122, 124,
              126–32
          division of labor, 126–28
          individual bee types in,
                  122–*23*
          origins, 130–32
          overwintering, 124, 130–31
          as "superorganisms," 118
      comb building, 119–*20*, 121
      communicative "dance," 138–*39*,
              140–43
      competition with solitary bees, 42
      defenses, 130
      economic value of, 105
      highly social nature, 44, 107, 143
      holometabolic development, 23
      molting of, 24
      navigation, 136–38
      nesting, 118–19, 119–*20*, 121,
              142
      orientation to sun, 134, 136–37,
              141, 143
      overwintering as colony, 124, 143
      and pheromones, 130
      pollen collection and transport,
              36–*37*, 38–*39*, 40,
              119, 127
      queen as regulatory mechanism,
              129–30
      races, in classification, 2, 116–17
      and royal jelly, 126
      sensory capabilities, 132–35
      swarms, 129–31, 142–43
      three types in colonies, 122–*23*,
              124–25
      trophallaxic communication of,
              128
      western, 115
  honeycombs. *See* bees: honeybees

bees (cont'd)
  Hylaeus (masked bee genus), 36,
          41–42, 45, 146
  in Hymenoptera order, 3–4
  jelly, royal, 126
  larva, 46, 54, 66, 72–74, 80, 99, 101,
          126–27
  Lasioglossum (genus), 66
      *Lasioglossum marginatum. See
              Evylaeus marginatus*
      microlepoides (sweat), 42
      pauxillum (sweat), 42
      zephyrum (sweat), 71
  leaf-cutting, 38, 41, 146
      female behavior, 50, 52
      number of species, 50
  Lestrimelitta, 97
  Lestrimelitta limao, 106
  Lithurgus (megachilid carpenter), 38
  Macropis labiata (mellitid), 62
  masked, 36, 40–42, 45–47, 48
      female behavior, 46, 48
      nests of, 46–47, 146
      number of species, 45–46
  mason, 38, 146
      bellflowers, 41
      blueweed, 41
      female behavior, 55–58, 59
      helicophile, 58
      nests of, 54–58
      species in U.S., 54
  mason bee of the walls, 50, 57–58,
          59, 79
      female behavior, 57–58, 59
  mating, 67–73, 75–77
      competition, 76
      pollination in, 76
      territories, 76–77
  Megachile (globally dispersed genus),
          50
  Megachile versicolor, 52–53
  Megachilidae (leaf-cutting; mason),
          38, 41, 50–51, 52–53,
          54–58, 81, 146

bees (cont'd)
  Melecta (carder), 81
  Meliitturga claviceps (panurgines), 50
  Meliponinae (stingless), 44, 83, 97
  Meliponini, 97–100, 102–6
  Melittidae (melittid), 40, 62
  melittids, 40–41
      flower preferences, 41, 62
      nests of, 62
  microlepoides (sweat), 5–7, 42–43 ,
          44, 65–72
  miner, 38, 40–41, 63
      nests of, 49–51, 146
  molting of, 24
  nectar, 115, 127–28, 134
  nest building, 21, 40–41, 46–47,
          48–50, 51–52, 53–58,
          59–62
      parasitic danger, 80–82
      protection, 77–78, 90–91, 121
      shared entrances, 65–66
      snail shell, 56–58
  nests of, 84–86, 108, 110–11,
          112–14
      human assistance to, 146–47,
              148–50
  Nomadinae, 59–60, 81
  orchid, 44, 73–74
  Osmia (genus), 38, 41, 54–58, 57,
          59, 146
      bicolor (mason heliophile), 56–58
      caementaria (mason), 59
      caerulescens (mason), 55
      conjuncta (snail-nesting mason),
              58
      mustelina-emarginata (mason),
              55–56
      papaveris (poppy mason), 56
      rufa (red mason), 55, 57
  overwintering, 64–65, 67, 70, 88, 96,
          117, 124, 150
  Oxaeidae (exotic), 44
  panurgines, 50
  Panurgus calcaratus (panurgine), 50, 65

INDEX

159

bees (cont'd)

Panurgus (miner), 38

parasitic, 80–82, 95–97, 106

in phylum *Arthropoda*, 3

plasterer, 40–41, 45–47, *48–49*

    nests of, *47–49*

    winterheath, 41

pollen, 18, 36–*37*, 38–*39*, 40, 46, 48–50, 60–*61*, 62, 90, 99, 121, 126

    basket collectors, 38, 40, 102, 122

    storers, 93

pollination, 76

predatory, 98

primevil, 6–7, 10

proboscises, 41–42, 59–60, 86, 92, 95

Psithyrus, 95–97

    sylvestris, 95

    vestalis, 95

pupa, 24, 46, 48, 54, 80, 92–93, 99, 126

queens, 32, 72, 86–*89*, 92, 96, 101–2, 119–*20*, 122–*23*, 124–*25*, 126–27

    emergency rearing, 128–29

    overwintering, 150

    and swarms, 129–32

reproduction, 46, 48, 52, 55–56

    See also under bees: nest building

resin (bee), 38, 55, 146

    buttercups, 41

royal jelly, 126

size variations, 108–*9*, 122–*23*

sleeping habits, 65

snail-nesting, *57–58*

social, 3, 13, 44

    semi-colonizing interactions, 62–75

    See also bees: colony building; honeybees

solitary, 3, 31, 35–36, 40, 42, 45–62, 83, 94–95, 99–100, *147*, 149

species, numbers of, 35, 44–45, 66, 80, 83

bees (cont'd)

Sphecodes albilabris (cuckoo), 81

Sphecodes (genus), 66

Stelis, 81

stingless, 32–33, 38, 40, 44, 76, 83, 97–*100*, 101–6, 121

    anatomical variations, 98, 102

    colonies of, 103–6

    defenses, 97, 105–6

    habitats of, 98

    nests of, 98–*100*

        involucrum, 98

    queen castes, 101

    trophic eggs, 102

    successive generations of, 69–75

sweat, 42, 66

    female nesting behavior, 65–72

    Fideliidae (exotic), 44

    fossils of, 5–7

    genera, *43–44*, 66

sweet. *See* bees: stingless

taxonomic classification, 42, 66

    difficulties of, 44

Trigona (genus), 83

Trigonini, 97, 102, 104–6

Triungulin, 79–80

trophallaxis, 128

tropical habitats, 35, 44, 50, 60, 83

vascular system, *19*

venom, 91

wasp-like predecessors, 8, 35–36

wild, 42, 45–63

Xylocopa (genus), 36, 41

    violacea (carpenter), 62

    Xylocopinae (carpenter), 36, 41–42, 60, 62, 64, 72

        Xylocopini (large carpenter), 72

beetles, 10, *13*, 29–31, 79–80

blister, 79–80

birds, 3, *13*

colonies of, 26–27

Bonn, Germany, 7

brachiopods, *13*

Brazil, 105

breeding, cooperative, 26–27
bumblebees, 7, 32, 38, 40, 44, 60, 76
Buttel-Reepen, H. von, 64
butterflies, *13*, 16
  metamorphosis in, 22, 24

C

Cambrian period, *13*
Canada, 9
Carboniferous period, 10, *13*
Carpathian Mountains, 117
castes, insect, 32
caterpillars, 22, 24
cephalopods (nautilus), *13*
Ceratina (dwarf carpenter), 64
cerumen, 98–99, 103
chitin, 15, 20, 22, 87, 129
chitons, *13*
chordates, 2–3, *13*
Chrysidadae, 78
class, genera, 2
classification, Linnaean, 1–2
coal deposits
  as source of fossilized life, 7
cockroaches, 9–10, *13*, 18, 22–*23*
coelenterates, 2
Coelioxys (cuckoo), 80–82
coelolepids, jawless, *13*
Collembola (springtails), 10, *13*
colonies, in nature, 27–28
  living structures, 33
conifers, 5–6, *13*
cooperative breeding, 26–27
corals, 2
crabs, 2, *13*
Cretaceous period, 5–6, 9, *13*
Cretatermes (termites), 9–10
crossopterygians, *13*

D

Dasypoda hirtipes, 78
Dermaptera (earwig order), 29

Devonian period, 10, *13*
dimorphism, 32
Dorylina (army ants), 3

E

earwigs, 29
echinoderms, 2
echinoderms (sea urchins), *13*
Eifel Mountains (Germany), 7
entomologists, 6
Eocene period, 7, *13*
Europe, 115
exoskeleton. *See* insects: anatomy

F

families, in nature, 28–30
family, within genera, 2
family associations, 28–30
  maternal, 31
ferns, 5, *13*
fish, 3, 29
  schools of, 26–27
flies, *13*, 148
Forficula auriculara (earwig), 29
Formicoidea (ants), 3, 24
fossils, 5
France, 117
Friese, Heinrich, 63
Frisch, Karl von, 138

G

genus (genera), 2
German roach, *23*
Germany, 7–8
ginkgos, 5, *13*
glaciation, Illinoian, 11
graptolites, *13*
grasshoppers, 18, 22, 29
group dynamics, 27
grubs, 22–*23*
gymnosperms, 5

H

hawkmoth, death's head, 121
Heterocephalus glaber (mole rat), 30
Himalayas, the, 110
Hodotermitidae (termite family), 10
holometabolism. *See under* insects:
  metamorphosis in
hominids, *13*
Homo erectus, 10–11, *13*
Homo habilis, 10, *13*
Homo sapiens, 11, *13*
honey, 107, 119, 121, 127, passim
honeycombs, 24, 33
hornets, 114
horsetails, tree-like, *13*
Hylaeus (masked bee genus), 36
Hymenoptera, 21, 122
  as bee order, 3–*4*
  family groups within. *See Apoidea*;
      Formicoidea; *Vespoidea*
  first appearance of, 9, *13*
  limited molting in, 24
  social structure in, 30–33

I

Ice Age, 10, 115, 117
Ichneumonids, 78–79
Illinoian glaciation, 11
imago, 22–*23*
Indonesia, 108
insects
  anatomy, 15–*17*, 18, 22
      See also bees: anatomy
  blood of, 18, 20
  classification of, 2–3
  colony-building, 24–26, 30–34
      social classifications, 34
  digestive systems, 18
  evolution of, 10, *13*
  larval stages, 22–*23*
  metamorphosis in, 22–*23*
  migrations, 27–28
  molting of, 22–*23*

insects (cont'd)
  nerve systems, 20
  origins of, 10–11
  reproductive systems, 21
  sensory powers of, 21–22
  vascular systems, 15, 18–*19*
instars. *See* insects: larval stages
Isoptera (termite order), 3, 9, 30, 33
Italian bee. *See* bees: *Apis: mellifera
      ligustica*

J

jellyfish, 2, *13*
Jurassic period, 9, *13*

L

larva, insect, 22–*23*, 24, 30
  parasitic, 78–80
  See also under bees: larva
Latin, in scientific naming, 2
Lebanon, 9
Leucopsis gigas, 79–80
Linnaeus, Carolus, 1, 115
lizards, *13*
locusts, 28

M

mammals, 3, *13*
  care of offspring, 28
  migrations of, 27
Meloe (blister beetle), 80–82
Mexico, 50
migrations, animal, 27–28
millipedes, *13*
Miocene period, 7, *13*
mites, 2
mollusks, 2
molting, 22, 24
mosquitoes, 26
moss, *13*
moths, 28

N

Necrophorus vespilloides (beetles), 30
nerve cord. *See* insects: nerve systems
New Jersey
    oldest ant fossils found in, 9
    oldest fossilized BEE found in, 6
New Zealand, 83

O

onychophores, *13*
order, genera, 2
Ordovician period, *13*
Orthoptera (primitve order), 29
ostracoderms, jawless, *13*

P

Paleocene period, *13*
paleontologists
    and origination of life forms, 5
Paleozoic period, 10
parthenogenesis, 26
Permian period, 10, *13*
pheromones, 130, 136
Philanthus triangulum, 79
Philippines, the, 108
phylum (phyla), 2
Pipa pipa. *See* Surinam toad, 29
plants, flowering
    coevolution with bees, 6
    origins in Cretaceous period, 5, *13*
Pleistocene period, 10, 117
Pliocene period, *13*
pollen, 5–6, 36–*37*, 38–*39*, 40–42, 46,
            48–50, 52, 56, 60, 62, 73–*74*,
            80, 127
Precambian period, *13*
propolis. *See* resin
Prosopis (masked species), 45
pterosaurs, *13*
Pterygota (insect wings), 10
pupae, 22–*23*, 24, 54, *74*, passim

Q

Quaternary period, 8, *13*
queens, *See* bees: queens

R

race
    in honeybee classification, 2, 116–17
reptiles, 3, 10, *13*
resin, 6, 9, 41, 73, 75, 90, 98, 103, 121
roach, German, *23*
Rott, Germany, 7

S

Sapygidae, 78
scorpians, sea, *13*
sea anemones, 2
sea urchins, 2, 13
seaweed, *13*
Seven Mountains (Germany), 7
shrews, 121
sigillaria, *13*
Silurian period, *13*
snails, *13*
sociobiolgists, 25
South America, 67, 106
species
    defined, 1
    as Linnaean concept, 1–2
Sphecidae (wasp family group), 8, 148
Sphecoidea (digger wasp group), 36
spiders, 2, 10, *13*, 148
    crab, 79
spiracles, *20*
sponges, 2, *13*
springtails, 10, *13*
starfish, 2, *13*
Surinam toad, 29
Swabia region, Germany, 7–8
Symphyta, 9
Systema Naturae (Linnaeus), 1

T

taxonomists
  and scientific classification, 2
termites, 22
  biparental families, 31
  colony-building, 10, 31
  origins of, 9–10, *13*
  and polymorphism, 32
  species of, 3
Tertiary period, 6–9, *13*
Thaumetopoeidae (moth family), 28
Thomisidae, 79
toad, Surinam, 29
Trechodes apiarius (beetle), 78
trees
  conifers, 5–6, *13*
  eucalyptus, 31
  squamaceous, *13*
Triassic period, 9, *13*
Trichodes apiarius (checkered beetle),
     78
trilobites, *13*
Triungulin, 79–80
tubules, nephridial, 18–*19*

U

Ural Mountains, 117

V

Vertebrata (subphylum), 2–3
vertebrates, classes of, 3
Vespa mandarina, 114
Vespoidea (yellow jackets, wasps), 3, 24
von Buttel-Reepen, H. *See* Buttel-
     Reepen, H. von
von Frisch. *See* Frisch, Karl von

W

wasps, 3, 32, 59
  capturing of prey, 8
  colony-building, 24, 33

wasps (cont'd)
  cuckoo, 78, 148
  diet of, 5
  nests, 33
  origins of, *13*
  parasitic, 78–79
  pupa, 24
  solitary forms, 31
  Sphecoidea (digger), 36, 148
  varieties of, 8–9, 36, 148
worms, 2, 10, *13*

Y

yellow jackets, 99

CPSIA information can be obtained at www.ICGtesting.com
Printed in the USA
LVOW090214060612

284862LV00001B/17/P

9 781441 929228